Diary of a Probie

DIARY OF A PROBIE

TURNING SETBACKS INTO STRENGTHS

NICHOLE O'DONNELL

FIRE ENGINEERING™
· BOOKS ·

Disclaimer
The recommendations, advice, descriptions, and methods in this book are presented solely for educational purposes. Photos are for instructional purposes only. Always wear the proper level of approved PPE when conducting training drills and operating at incidents. The author and publisher assume no liability whatsoever for any loss or damage that results from the use of any of the material in this book. Use of the material in this book is solely at the risk of the user.

Fire Engineering Books
110 S. Hartford Ave., Suite 200
Tulsa, Oklahoma 74120 USA

800.752.9764
+1.918.831.9421
info@fireengineeringbooks.com
www.FireEngineeringBooks.com

Executive Vice President: Eric Schlett
Vice President, Group Publishing: Amanda Champion
Vice President of Content Operations: Starlet Franz
Sales and Customer Service Manager: Lane Nash
Managing Editors: Diane Rothschild and David Rhodes
Production Manager: Tony Quinn
Senior Book Development Editor: Daniel Edward Petrino
Book Designer: Robert Kern, TIPS Publishing Services, Carrboro, NC
Cover Designer: Brandon Ash

Library of Congress Cataloging-in-Publication Data Available on Request

ISBN: 9781593705268

Printed in the United States of America

1 2 3 4 5 29 28 27 26

CONTENTS

FOREWORD

I knew Nichole for a short time before she came to the fire service, as a town employee for the highway department. I heard the stories of her being a tough girl—an arborist running crews—and I thought to myself, *Hmm . . . sounds like she is up for the job, but of course I will be tougher.*

Always in a good mood and instilling the same positivity in others, Nichole strolled into the firehouse full of joy and ready for the next chapter in her life. At the time, I told myself she was just another probie that we were going to have to babysit and teach the ropes, but Nichole was focused and ready to be the best firefighter that we had ever had in the department. This was her curse and her salvation.

Her journey since has taken many turns, and it has been full of ups and downs. This job can take everything from you, but it can also give you everything that you could ever need. Nichole had to overcome many personal and professional challenges to get through her first year as a probationary firefighter. I am extremely proud of her accomplishments, and I am glad that I was able to be part of the journey. *Diary of a Probie* instills in the reader that it is never too late to heal yourself and come out on top. Her never-give-up attitude is contagious and will make you work harder than you ever thought you could. All I can say is, "Just keep being you, and you will succeed!"

—Anthony Arventos, captain/paramedic,
Wilbraham (MA) Fire Department

PREFACE

Hi there, whoever you are. Whatever brought you to this book, I'm glad you're here. It doesn't have tales of heroics, firefighting tips, or a step-by-step guide to saving the world. It's a glimpse into real life, told through the lens of someone who found purpose in the fire service and stumbled through chaos and failures to get there.

This book is for anyone who's ever chased a dream, questioned their worth, or tried to find solid footing in a world that doesn't slow down. You'll read about the times I fell down the stairs at an electrical fire, flooded the kitchen with coffee, and gave the whole town a strip show when I split my pants dressing a fire hydrant. And then there are the "Dear Chief" moments when I had to sideline my pride admit my own harsh mistakes. These stories don't make the news, but they do shape your character, and if you can learn to laugh through them, you're already halfway there.

At heart, these stories are about something beyond the uniform. They are about learning how to show up when everything in you wants to quit. They are about the people who believe in you before you believe in yourself. I didn't have a clean or easy start. I made choices when I was younger that should've taken me out of the game altogether, and the odds were against me on the job and in life for a very long time. In my early 20s, I turned it all around, becoming a firefighter, author, and aspiring paramedic.

Whether you're in the fire service, navigating your own career path, or just flipping pages looking for something real, I hope you find a piece of yourself in these true stories, especially the raw, honest, and vulnerable ones. Pull up a chair. Grab a cup of coffee, but make sure I'm not the one pouring it. Welcome to the journey—the raw, unfiltered ride through the year that changed everything. I should've been another name on a headstone or another mugshot in the system—but instead I'm here, torch in hand, and this book is me throwing the match into your gasoline. The weak watch. The driven ignite.

ACKNOWLEDGMENTS

I wouldn't be where I am today without the unwavering support of so many incredible people. First and foremost, I owe endless gratitude to my best friend, Alycia. Her sharp eye and no-nonsense feedback on early drafts were invaluable. She's been the support I never knew I needed. Her great advice, friendship, and guidance—and being there not because she had to but because she wanted to—have meant an incredible amount to me. She's the once-in-a-lifetime kind of person that I'm lucky enough to call not only my best friend but family. She's the best family I've ever had—the most supportive, caring, beautiful human being.

That said, thanks to my "niece and nephews" (Gianna, Vincenzo, and Tommy), for including me in this crazy family of yours. It's an honor to be called Auntie Nichole. I'm proud of the three of you and always will be, so keep kicking butt and taking names. You're all going so far, and I can't wait to watch you succeed in everything you set your minds to. And Tom, thank you for accepting me into your family and dealing with my outbursts. Thanks for trusting me with the kids, giving me advice when I needed it the most, and helping when I was down. You may come off like a grizzly bear, but you do have a strong heart deep down, and I'm glad to have witnessed that side of you.

To my brother from another mother Nick "Nicky" Corigliano, you were my rock through this initial process. You are not just family, but my mentor. As lieutenant, you knew exactly what it took to make it. You thought of everything—from edits to my resume and interview preparation advice to reminders about sending a thank-you email afterward. For all that you did, I'm forever grateful. You set me up for success in the long run with the best job in the world. You've done more for me than you'll ever know. Love you, Nicky—thanks for taking a chance and believing in me.

To my friend, mentor, and fellow firefighter/paramedic Adam Martin, thanks for being the guinea pig who read the first draft of this book. It all started as an idea during our Public Fire Life Safety Educator course back in May 2024, initially called *The Probie Bible*. Seeing the potential for this book, you told me, "I'd read that in a heartbeat—it's really good!" If you hadn't told me to go for it, I wouldn't have submitted my proposal to Fire Engineering Books, and this book wouldn't exist. Adam, thank you for being such a great friend. Your support is valued, and you have a special place in my heart. I am glad you have been along for this entire journey.

To my academy instructors—Mr. Rankins, Mr. Rhodes, Mr. Cross, Mr. Clegg, Mr. Schuetze, Mr. Hatch, Mr. Boucher, Mr. Grimaldi, and Mr. Girard—who I'm honored to now call my friends, thank you for your high standards, wisdom, and relentless commitment to excellence. The lessons you instilled in me will last a lifetime. Your leadership and dedication pushed me to be my best, even when it felt impossible.

A special thank-you to Captain Tony Santinello for being one of the instructors who has been here since I began in this field in 2021. He taught me one of the coolest clove hitch tricks and came in early every academy day (on his own personal time without getting paid) until I understood hydraulics. He has helped me grow as a firefighter and as a brand-new recruit. I wouldn't have such a solid foundation with pumping operations if it wasn't for him. Over the past 4 years, he has become a friend and mentor, and I'll always appreciate what he did for me as an up-and-coming firefighter.

To my mentor, family friend, and teacher District Chief Robert McCaffrey, thank you for giving me the chance to prove myself and putting in a good word for me at my career department. You gave me an opportunity, and I ran full force with it.

To everyone who provided feedback and shared advice over a year of writing, I can't thank you enough for your wisdom and guidance. You've shaped this book in so many ways.

To Steve Makos, my friend, sounding board, and a 25-year veteran firefighter/paramedic, you define what this job is all about. From the moment you walked up to me at the academy in 2021 to that first motorcycle ride, you've been a constant in my life. You've checked in, kept me honest, and reminded me never to lose sight of who I am or why I'm here. Your idea for the signature and notes pages gave this book its soul, but it's your testimony and support as a friend that left the deepest mark. I'm proud of you—not just for 25 years on the job, but for the way you've carried yourself through it all. You're someone I'll always look up to—in the firehouse and in life.

To Fire Engineering Books and the entire team behind the process, thank you for believing in me and for your trust in my work as a 23-year-old inexperienced probie. I am honored to partner with you on this journey.

To my mentor and friend who I call family Joshua "Sterls" Sterling, you saved me when I needed it most. Thank you for being a call away. Your unwavering support has been a catalyst to my success in the fire service from day one. You are the reason why I applied to my career department now; you saw something in me that I didn't see in myself. I would not be writing this acknowledgment for my first book if it wasn't for you. There is no such thing as a self-made individual—you are part of my journey and success.

To my friend, mentor, and officer Patrick Farrow, thank you for everything. I'll never forget your encouragement and wisdom. As a fire officer, you help people turn this into a career in public service. I appreciate that your goals have always been to keep me safe and help me succeed.

To my call department chief Poulin, thank you for giving me the opportunity and providing the foundation for the career I've built today. I wouldn't be where I am today without you.

To my peers at the Hampden (MA) Fire Department, thank you for being part of every step in my journey as I grew and evolved in this field.

To my peers of the Wilbraham (MA) Fire Department, thank you for the guidance and support you've given me along the way. I know it was a leap of faith taking a chance on a probie with limited experience like myself, but I've done everything I can to learn, grow, and earn my place among you. It means a lot to be respected on this team. Thanks for making me a part of the family, even if it took me a bit longer than others.

To Christopher Ott, thanks for believing in me and helping me get my fitness on track to become an asset as a firefighter. You've been a friend when I needed one the most, always supporting the decisions I made and checking in on me throughout the process.

To my coach, friend, and mentor Scott Barger, thank you for being my rock in the fitness and fire world, turning me into the ultimate asset by helping me lose 30 pounds. You've been one of my biggest supporters and a friend when I needed one the most. You have done so much for me in the short amount of time I've known you, and you've become family. You're going to do everything you set your mind out to do—I can't wait to see your business take off. Meanwhile, I got you when it comes to drafting. Proud of you!

To my career department chief Andrews, thank you for believing in me and giving me a lifetime opportunity to prove myself. You trusted me to grow into an asset, even though I came in with zero experience. That trust means everything.

To Captain Anthony Arventos, I want to acknowledge the profound impact you've had on me not just as a firefighter but as a person. From the moment I walked in those doors for my very first interview, you made me welcome, and it felt as if I had known you for years. Once I was appointed for the position, you were the first person to come up to me, look me in the eyes, shake my hand, and congratulate me. When I started out, I thought this job was all about skill, endurance, and grit. But you showed me it's so much more.

Having a great leader doesn't matter only in your first year; it matters every single day. During that rough patch in April 2024 when I hit burnout, you sat me down in your office for a disciplinary meeting, and I was terrified—not of you but of admitting that I was struggling and that I was failing myself. I thought you'd look at me differently after that, but instead of judgment, I was met with understanding.

You recognized the walls I had built—I thought they would protect me, but they were suffocating me. When I couldn't see a way out, you guided me, reminding me that this job is brutal, but I'm tougher. That conversation was about everything I'd been carrying for so long, the pain and trauma I'd buried. You didn't have to help me. You could've given me the usual advice to toughen up or to not let it get to me. Instead you told me the truth: "If you keep bottling everything up, pretending to be fine, this job will destroy you."

You didn't sugarcoat anything, but you never made me feel weak either. You made me feel seen and let me know that it was OK to be human and not have all the answers right away. Your words of wisdom helped me find balance. I listened because you knew a thing or two, and I appreciate that you've always kept it real with me.

Walking in as a probie to a department full of seasoned firefighter/paramedics, I often wondered if I was really cut out for this job. You reminded me that no one becomes experienced overnight, that even the best ones stood exactly where I was. You told me to stop comparing myself to everyone else and to focus on getting a little better every single day. You saw something in me that I couldn't yet see in myself. You've been like a dad to me in this career—letting me make my mistakes, but never letting me fail. I'll never take that for granted. The respect, trust, and patience you've shown me is more than I could've asked for as a probie. You're the reason I look forward to coming on shift. From my first day, you made me feel like I belonged. You never give up on me, talk down to me, or make me feel small. When you pick on me, it always makes me smile, and I appreciate that you look out for me in my not-so-finest moments.

You've shown me what it means to be a trusted leader, never letting gossip get in the way of being real and standing up for me when it wasn't easy. The odds were against me, and never once did you let anything stop you from believing in me. From the beginning, you knew I was going places, and I'll take that to the grave with me. Somehow you never lost your patience with me. Yes I know I'm more work than your actual job, but hey, that's what happens when you're a good leader and you leave a mark on someone like me! Your honesty in one of the lowest moments in my life gave me the push to find an outlet, and that's when I began writing. What began as an escape has now become this book. You have been the catalyst for my success in the fire service, even though you modestly say, "It wasn't me. You did the work."

(Side note: Sorry for almost killing you while driving. You did have nine lives, but now you're down to eight thanks to Nicholey! Yeah, I definitely won't do that again. And by the way, when I'm famous, I'll make sure you get your cut.)

I'm the firefighter I am today because of you. And when the day you hang up your helmet finally comes, it's going to break me a little inside. You've been more than a captain to me—you've been the voice that pushed me when I doubted myself, the steady hand that never let me fall, and the reason I found my place in this job.

When you step down, I'll carry you with me into every fire, every call, every decision I make. Not as just another lesson learned, but as a part of me. No matter where this career takes me, I'll always be chasing the example you set, and I'll never stop trying to make you proud.

From this day forward, I'll lead others the way you led me, ensuring your legacy never leaves the firehouse.

Thank you to everyone who has walked this path with me, offering wisdom, advice, or just a kind word when I needed it most. This book would not have been possible without all of your unwavering support. There are so many people who have helped me from the very beginning and deserve credit. My gratitude extends to all of you.

To my friend and mentor who changed my life, Payton Owens, captain of the DeKalb County (GA) Fire Department and president of the Metro Atlanta Fraternal Order of Leatherheads Society (FOOLS), thank you for your part in the one video that changed my life. It all started when I tagged you on a video post in early January 2025. I didn't think much of it at the time—but it took off. That post went viral, opening the door for me to connect with you and the Metro Atlanta FOOLS. Your unexpected invitation to join flipped a switch. Suddenly, I was part of this tight-knit, ride-or-die crew—the ultimate underdog fire group. We weren't the biggest, but we were the most real, the most raw, the most relentlessly supportive—and it showed. From that moment, my social media posts started gaining serious traction. We didn't just ride the algorithm—we became the algorithm. Fire reels, collaborative posts, shoutouts, shares—we were everywhere. And before I knew it, my name held weight in the fire service, with over 10,000 followers on Instagram. Payton, you told me early on that I needed a catchy handle for my Instagram page and a logo to brand myself, something people would instantly recognize and associate with me. From day one, you saw potential in me before I even saw it in myself.

You pushed me to dream bigger, and because of your belief in my capabilities, I leaned into my vision for *The Salty Rookie* (shoutout to Kyle Lawrence who created my logo). Thank you for being the reason I started my branding journey. Your guidance helped me grow both online and in the fire service. I'm proud of everything you've overcome; you've truly inspired me. And thanks for idea for *The Kitchen Table Talk*, which I turned into my podcast.

To my brothers of Metro Atlanta FOOLS—thank you. You've shaped this journey more than you know. When no one else was paying attention, you were the ones sharing my content, reposting, commenting, collaborating, and encouraging me. You didn't just support me—you amplified me, and that's something I'll never forget.

I'm proud to know each of you personally, to stand beside you, and to call you brothers. Each one of you represents the standard I admire most in the fire service—grit, humility, and heart. Thank you for believing in the mission before the rest of the world saw it and for proving that brotherhood still runs deep.

This book may carry my name on the cover, but it carries your fingerprints on every page.

To Juan, the owner of Working Fire Company, thank you for believing in me since day one. You didn't just help launch my merch line, you backed my vision before it had momentum. Your constant support and encouragement have meant more than words can express. It's rare to find a company that not only represents the fire service but also lifts up the individuals within it. Thank you for standing beside me on this journey and helping me bring my brand to life.

To Ben and Sam from Back of the Bay, from the very beginning you two have been a source of motivation, integrity, and truth. You two are the iconic faces of doing the right thing, even when no one's watching. You stand for what matters. You speak up, show up, and never stop grinding. Your relentless drive and commitment to bettering the fire service has inspired me in more ways than you know. I'm so thankful you've been part of this journey with me. You set the bar high, and you do it the right way, the only way. In your famous words, "DO BETTER."

A further thank-you also goes to all of my haters who dragged my name into the mud even while I wasn't there to respond. Your doubt and criticism lit the fire under me to tell my story of success and coming out stronger on the other side. Because of you, I leveled up and became the best version of myself, and for that motivation, you have my appreciation.

Last, I couldn't have done any of this without God. The truth is, I shouldn't even be standing here telling my story—I should have been dead or in jail a long time ago. But God had other plans. He carried me when I felt I didn't deserve it, protected me when I didn't even realize it, and showed me grace I could never earn. Every open door, every setback, every heartbreak, and every victory—it was never chance, it was Him.

This book may carry my name, but every page belongs to Him. It's my story, but it's His testimony. And as much as I've fought, endured, and overcome, it's only because He was with me every step of the way.

So if you take nothing else from these pages, take this: With God, nothing is impossible.

> *For I know the plans I have for you," declares the Lord, "plans to prosper you and not to harm you, plans to give you hope and a future.*
>
> *—Jeremiah 29:11*

I am living proof of that promise.

INTRODUCTION

Don't Short Yourself

In March 2024, I applied to a local paramedic program, which was a job requirement for my department. There was a catch with the application, though: a prerequisite English placement test to prove if you were competent or illiterate. My heart rate spiked, and anxiety kicked in full throttle. I was immediately concerned with the requirement because I had struggled early on to overcome learning disabilities, attention deficit hyperactivity disorder, and reading obsessive-compulsive disorder (which is when you read the same text five times and still don't retain the information). I was the kid who couldn't sit still or focus, who took three times longer than everyone else to learn.

I walked into that examination fearing that I'd prove the doubters correct, thinking to myself, *There is no way I'm going to pass. I took summer school 3 years in a row and barely graduated. How am I going to pass this stupid test?*

The first half of the cognitive examination was multiple choice, and the second half was a writing comprehension assessment. I embraced the challenge, despite being filled with doubts. The essay was a response to the question "Is it better to be a realistic or optimistic person?" Feeling the odds were stacked against me, success was not my expectation, but somehow words began flooding the paper. I realized I wasn't writing to define my comprehension level; instead, the moment was a revelation of my destined career as an author, which felt like something I was meant to do all along. I clicked submit on the examination and said a prayer.

Once the results were posted, I read my feedback with amazement: "In all my years here, this is one of the highest scores I've seen, and your writing is beautiful. I want to read more!" That's when it clicked: I had found my niche, which utterly blindsided me. I was a lot smarter than I had realized, and I didn't need a college degree to be successful. A passion to write had made its way to the surface, and the words of that essay planted the seed that has grown into *The Diary of a Probie.*

How This Book Came to Be

A senior firefighter once told me, "Write everything down. Moments you think you'll remember because one day you won't." So I kept a journal, and the storytelling began. Within 6 months, I wrote an entire book filled with my probie experiences, with zero intention that anything would come of it. I shared unique moments that shaped me into the firefighter I am now, and eventually it turned into something I had never imagined—my first book.

When I learned of Fire Engineering Books, I realized I had a chance at becoming an author. Seeing the open doors of opportunity, I took a leap of faith and sent in my proposal. Even though I didn't have decades to earn credibility, I went for it anyway. *What's the worst that can happen? I have nothing to lose.* Never did I think that the most credible company in the fire service would consider publishing someone like me, still in my early 20s and essentially fresh meat in the fire service. Faith fully took the reins!

I didn't have a background of generational firefighting experience. I was navigating an entirely new world and quickly realized how overwhelming this career can be for anyone. What started out as simply a journal in which I would write everything down turned into my way of giving back. I hope this will be a tool that helps probies to thrive in the fire service—and individuals to thrive in life.

This is a tribute to my first year in the fire service: the chaos, the breakthroughs, the people who had my back, and the ones who made me fight for every inch of ground. Every mistake, every win, every moment that tested whether I had what it takes, and every one that proved I do. I didn't come from a polished past: I grew up being told I wasn't smart enough, strong enough, or built for much of anything. I dragged myself up through the mud, chasing approval that never came. That ends here: I found my voice. I found my fire. And I finally stopped asking for permission to use either.

Welcome to my journey—a raw, unfiltered ride through the year that changed everything. The fire that nearly burnt me out is the same one I now carry as a torch. If it lights something in you—good.

Directions for the Interactive Pages

I'm big on taking notes and making sure I've got something I can hold onto, something I can look back on. I wanted to give you the same opportunity, so I came up with a book that you can interact with, one that's truly yours! Each chapter concludes with prompts to reflect on your own experiences. This is your space, so take it and run with it.

There's also a signature page at the end of the book—like the end of a high school yearbook but way cooler. This is where your academy instructors and anyone who gets to witness your progress firsthand can leave their mark, comment on your performance, or write a customized note. Firehouse members can write a few words about your probie year, which is going to fly by—before you know it, all those little moments will start to blur. This is all about capturing the experience while it's fresh, giving you something to reflect back on years down the road. You only get one shot at your probie year, so make it one to remember.

1

INTO THE REAL WORLD

September 19, 2023

I learned the meaning of "nauseous," and I wasn't alone, together with 35 recruits, battling through 8 hours of hell: Recruit Fitness Orientation (RFO). Stepping into the drill yard and immediately wondering, *What did I get myself into*?

The day was a brutal mid-80s, unseasonably warm for Massachusetts in September. The kind of heat where the birds stop chirping, the air is thick with silence, and you can hear a bucket loader miles away. The day began with orders from seasoned instructors who looked like they ate nails for breakfast and washed them down with gasoline. They barked professionalism, but their directions were a blur that made it clear that failure was their expectation if you were weak.

The fear was real, lodged in our throats like a lump of lead. The air was filled with groans and wheezes from brand-new recruits who woke up one day with the idea to be a firefighter. You could tell who was in it for the job or just for the badge and the cool T-shirt. RFO was a meat grinder, and it didn't take long to pick off the weak minded.

Twenty different voices hammered at once:

> *"Come on! That's all you got?"*
> *"I guess this job isn't for you!"*
> *"The door is right there. No one's holding a gun to your head!"*
> *"Faster! Quicker!"*
> *"My grandmother could do better than you!"*
> *"Not impressive."*
> *"You call that work ethic?"*

Amid chaos and demands, there I was: a completely lost probie trying to focus on the task at hand while pushing through what felt like the longest minutes of my life. Feet trembling raggedly, legs barely holding up, asking yourself *When will it be over*? and *Am I really cut for this*?

We started with a 1-mile run at an 8-minute pace set by the instructors. For someone like me who hates running, that was a sprint. But you push through no matter what that little voice in your head is screaming at you. The discomfort was excruciating. I wanted to puke, and some recruits were dry heaving or losing their breakfast right in the drill yard.

Stopping your assigned task meant a trip to emergency medical services (EMS), and going to EMS meant you were out of the program. Words of encouragement passed among us, recruits trying to lift each other up. My partner from my department slowed his pace to run with me, telling me, "One foot in front of the other. You've got this. We're almost done!" That kept me going. Without him, I would've listened to that voice telling me to quit.

After the run, we had 1 minute of nonstop work at 8 stations: bear crawls, hose drags, hose pulls, push-ups, dips, step-ups, burpees, and tire slams. The hard work continued, and by the end of it, everyone was gassed.

An hour into the workout, we got a break. I grabbed water, thinking we were done. Nope—this was just the beginning. Note to self: Never make assumptions in the fire academy, much less on the fireground.

"Time to step it up." The look on everyone's faces was like a deer in the headlights. "Next round is 45 seconds of work, run a lap, and move on to the next station." Just when you thought you couldn't take any more, after completing all stations at 45-second intervals, we had to do 30-second rounds, running between stations. Drenched in sweat, faces beet red, we looked like we'd been tossed into a pool fully clothed. It was gross and uncomfortable, but we made it through, chanting, "Embrace the suck!"

The academy was a test of endurance—physical and mental. By the end, we were down to 27 recruits from the 35 who started; 8 had been sent home with their hearts broken. A little more effort could've saved them, but it just wasn't in them. Effort goes a long way, and any instructor will respect you for it.

Key Takeaways

Stop waiting to feel ready. You don't need to be fearless to begin; you just need to begin. Show up for yourself, even when it's hard. Especially when it's hard. Because the fire service doesn't just change your career; it changes who you are at your core. And once you see what you're capable of, there's no going back.

You're not just earning a badge. You're earning a new standard for yourself—one that conveys, "I don't back down. I show up. I lead. I overcome." And that mindset? That's forever.

Final Thoughts

Pain is temporary, but this job is forever. Ten weeks for the best career of your life—that was the mantra we held onto. My mind screamed at me to quit, to stay comfortable by tapping out. I almost listened. I almost convinced myself that an average life was enough. But this job doesn't reward comfort. Instead, it demands grit: You must show up when you're exhausted, act when you're afraid, and think clearly when chaos erupts around you.

This experience taught me what I was made of. It taught me discipline, humility, resilience, and that growth doesn't happen when things are easy. Growth happens when your body is broken down, your mind is screaming, and you decide to keep going anyway. Because someone out there is going to need you on the worst day of their life, and when that day comes, you better be ready.

Every time you do something that drags you out of your comfort zone, you evolve. You level up. I let a poor mindset hold me back for years, telling myself I wasn't smart enough, strong enough, or cut out for more. But knowing what I know now, I'd relive that first brutal day of RFO again in a heartbeat; I'd face it head on and crush it, even with fear and doubt whispering in my ear. It was never the pain that held me back; it was my mindset. Once that changed, everything else got better (fig. 1–1).

FIG. 1–1. Massachusetts Firefighting Academy graduation, January 27, 2024 (Photograph courtesy of Jared Berg)

Reflection Prompts

✠ What are the requirements to qualify for your local fire academy?

✠ What parts of the fire academy did you find most difficult, and how did you push through them?

2

WELCOME TO YOUR REALITY

The Massachusetts Firefighting Academy is a whole different breed, having a duration of only 10 weeks. This may not sound incredibly demanding, but the pace is relentless. If you think you want a short academy, think again. Knowing what I do now, I wish our program was longer, for the benefit of both instructors and recruits.

An incredible amount of information and training is crammed into only 50 days to meet national requirements. For example, Rapid Intervention Training, which should be at least a week, was crammed into 2 days. We come out hoping to retain enough of the lesson to be able to apply it on the fireground.

While I'm not an expert on other states' requirements, most academies stretch their programs out, giving recruits more time to absorb the material and get comfortable with the physical demands. Here, though, it's a race against the clock, preparing for Pro Board certifications, drill yard practicals, chaos, New England firefighting, and of course, ladders.

We drilled ladders repeatedly, until we got it right. Workout over? Time for ladder drill! Climb, lift, carry. I'd rather run a mile than ever do another ladder drill at the academy.

Endurance and Functional Fitness Focus

Minimal effort? Think again! You have to be able to run a mile in 8 minutes to survive orientation. Get familiar with workout styles like AMRAP (as many rounds as possible) and EMOM (every minute on the minute), which are your companions not only for the duration of the academy but throughout your career. It's not just about brute strength; you need stamina to make it through. You won't see results right away, but patience is a virtue. Investing in yourself will pay off. By the time you reach the end of the program, the improvement in your performance will be staggering.

Prepare yourself to succeed in the academy. These rules can be applied on the job as well:

1. Show up early every day.
2. Have a strong foundation for a fitness regimen going into the academy.
3. Review your material before class starts.
4. Take time to check your gear. Make sure that everything's compliant and nothing's missing.
5. Don't complain.

6. Follow orders.
7. Don't wait to be asked. Jump in where help is needed. Volunteer to go first. Show that you care about the job.
8. Keep up with good hygiene. Don't be the one who needs to be told to shower because you smell.
9. Hold yourself accountable.
10. Be prepared to hear the voice that tells you to quit when it gets uncomfortable. Work through it.
11. Show up with a positive attitude, and address everyone as sir or ma'am. No names. Remember: You're a recruit. Even if you know a staff member, treat them as an instructor, and don't refer to them by rank or name.
12. Don't be arrogant. Word travels fast. If you act up, everyone will know, and not in a good way.
13. Stay out of trouble. This means no alcohol. Yes, it'll be tempting. Your class becomes your family—all struggling together, building a bond that's hard to explain to anyone outside of the academy. So when the weekend comes, there's this pull to blow off steam together, to go out, laugh, and feel a little normal after the grind of the week. However, this isn't college; it's your career, and the stakes are higher. That one drink or that one night out could lead to a reputation that sticks with you long after the academy. Right now, you're setting the foundation for who you'll be in this field. Protect that fiercely. If you ignore this advice, don't say I didn't warn you.

Instructor Schuetze's Orders

Instructor Schuetze, our lead instructor throughout the academy, quickly became a personal friend. These were his top three expectations at the academy:

- *Don't eat more than your mass can hold.* Wise advice, but one of our recruits didn't get the memo. He had the great idea to chug a full gallon of milk at lunch, paired with spaghetti, meatballs, and acid reflux–inducing sides. With his belly full of carbohydrates and whole milk, this guy was struggling. Sitting in our end-of-day lecture, suddenly he bolted up into a full sprint stride. He didn't make it. Before he could reach the trash can, he explosively hurled colors of the rainbow, covering desks, students, carpets, and my hair—an involuntary cut and color. Mr. Schuetze reminded him, "This is why we told you 6 weeks ago to never eat more than your mass can hold. This is what happens." Despite that rough start, he drank almost a gallon of whole milk each day, throughout the academy. Personally, I could never stomach milk on a hot day, let alone while grinding through 8 hours of brutal work, 5 days a week. Coming from experience, I don't recommend milk as a hydration choice.
- *Don't cheat on your reps.* Chest to deck! No half-ass reps! No one's going to hold your hand.
- *Just get it done.* Ten weeks for the best job in the world.

Academy Expectations

Our 10 weeks were jam-packed with drills: confined-space entry, search and rescue, victim recovery, infant recovery, ladder climbing, grueling workouts, and live fire training. This was nothing I hadn't prepared for; I had worked so hard, and now it was time to prove it. In the academy, the fires we faced were made of hay bales, straw, and pallets—all clean-burning materials.

That's the only time we ever experience that kind of "safe" fire. Once we graduate, reality sets in. Real structure fires are filled with carcinogens and various chemicals that can kill you in an instant if you breathe them in. That's why using self-contained breathing apparatus (SCBA) is so important. This one piece of equipment stands between us and the deadly toxic substances we face on the job.

I thrived on being thrown into scenarios I'd never faced before. For example, long lug out was a great drill: You're placed in a room in blacked-out conditions. Disoriented, you have to make a Mayday call and wait to be acknowledged. Mayday calls are life-or-death situations; if you hear a Mayday, all radio traffic ceases, as a firefighter attempts to get their distress call through.

Training on the Mayday process uses the mnemonic LIPS (location, identification, problem, solution). It sounds like this: "Mayday, Mayday, Mayday"; you pause, wait for acknowledgment, and once received, transmit your message: "Third division, Charlie corner near the window, Firefighter O'Donnell, lost and disoriented, attempting to find my way out." Then you manually activate your personal alert safety system (PASS) device, which helps the rapid intervention team locate you. Finally, you find the hoseline and follow it out.

The Trick to Finding Your Way Out

Smooth-bump-bump to pump. When you feel the hose coupling, moving from the smooth to the bumps indicates that you're heading toward the pump and safety. If you feel bump-bump-smooth, you're heading the wrong way, back toward the nozzle and deeper into danger. When you make it out, you radio back to cancel your Mayday and let the incident command know you're safe.

If you relocate divisions while you're lost and disoriented, you need to notify command, because if the stairwell suddenly collapses, they need to locate where you are. Time is precious in those moments, and your air won't last forever.

These drills tested us in the most uncomfortable situations. That's what I love about them. They put us into real-life scenarios, including potentially fatal situations.

Need-to-Knows

You receive assignments 6 weeks before you even step foot on campus (at least in Massachusetts), and you better start studying beforehand. Don't presume that you can get by on common sense! The written examinations can trip you up just as easily as the physical challenges. These vary across the topics in Firefighter 1 and Firefighter 2—that is, the basics of firefighting—all of which end up meshing together.

Stay on top of your studies from day one. If your academy includes the hazmat examination, buckle up. It's a grueling 40 hours (varying by state or district), but an essential foundation for your career. That certification lasts a lifetime—and is a prerequisite for promotions, job opportunities, and academy classes.

Firefighter 1 consists of the minimum training requirement to make entry on a burning structure. You learn about hoses, building construction, ladders, packing, deploying, forcible entry, rescuing victims, fire behavior and extinguishment, fireground communication, and salvage and overhaul, among other fundamentals. Importantly, though, in Firefighter 1, you aren't supposed to enter without trainees in Firefighter 2.

Firefighter 2 encompasses both ancillary training and being proficient in Firefighter 1 tasks. The additional learning includes preplanning, building inspections, public education, fire investigation (you don't actually investigate, just note details to tell the fire investigator and how to properly maintain a chain of evidence, either toward the point of origin or potential arson), and sprinkler systems.

Our examinations were weekly, pertaining to what we read and not so much on the physical aspect in the drill yard. Anything in the textbook was fair game to be tested on. We also had practice quizzes available on our online portal for the academy, and any quiz question was fair game on the examinations.

If you did the practice quizzes, you'd pass the examinations with no issue, although you still had to read the required textbook material to do better than 90%. However, if you didn't do the work, you'd fail your weekly examination. Until we got to hydraulics, I hated math; the assigned problems were complex scenarios (that you're probably never going to be in), but if you could solve those, then you'd be able to do any fireground hydraulics calculations with no issue.

For the Firefighter 1 and 2 examinations, invest in a good examination preparation book. By making this investment, I ended up scoring an 87%, which was a big win considering my history with test taking. Passing that examination was a testament to how hard work can pay off if you apply yourself.

After the Academy

Ten weeks of pure hell for a 34-year career—that's what the instructors drilled into everyone. Once the academy is over, you'll miss it, no matter how tough it was. It's bittersweet, but the lessons stick with you for life. You learn how to survive, keeping yourself and your crew safe. They won't let you fail, as long as you're willing to adapt, learn, and push yourself.

The academy is your foundation, but the real challenge starts when you hit the field. If you've made it through a full-time fire academy, that doesn't mean you have what it takes. The goal once you arrive on shift at the firehouse is to prove yourself to your peers—and most importantly to yourself. Yes, it's an everyday challenge, worth the difficulties you face because the reward outweighs the setbacks.

Now, it's about keeping that momentum, staying hungry, and striving to be better. You can't afford complacency. An online search using the phrase "firefighter complacency" will show you why. Every day brings new challenges, new lessons, and new opportunities to grow. So show up ready to work, take care of your body and your mind, and hold yourself accountable. Your performance in the academy reflects on your department, and in the field, it could mean the difference between life and death.

Key Takeaways

Having a strong foundation is number one. Ninety-nine percent of success comes from showing up early and not having a bad attitude. If you care enough to put forth the effort in the academy, that will follow you through a strong probationary year. It's not going to be easy. However, the end reward of the best job in the world is worth it.

You can go as far as you want in this career. There aren't any limitations if you're passionate enough. The academy sucks sometimes. That's the truth, but here's the trick: Don't just survive the suck; flip it into a challenge you're hellbent on beating! Every miserable moment is an opportunity to prove you're tougher than the discomfort. It won't last forever—grind, push, cry, and bleed if you have to; just don't quit.

Final Thoughts

A strong first year will build a lasting foundation that'll help you to succeed in the long run. As someone who has navigated this field without a clear guide, I understand that it's not going to be easy, but the reward will be the victory of a lifetime. It's not about how many times you fail, it's about how many times you get back up when you're knocked down without letting challenges overpower you. Your mindset can hurt you or help you. Make mistakes and learn from them. You've got this!

Reflection Prompts

- If you're currently in or have completed the fire academy, how did you prepare for it, and did you feel ready? If you're not in an academy yet, how are you preparing for it?

- What's the most important piece of advice you'd share with someone interested in applying to the fire academy?

3

UNDERDOG TO UNSTOPPABLE

August 28, 2023

Tonight I was appointed by the selectboard to the career firefighter position I'd always dreamed of. As I stood there, I couldn't help but reflect on the path that led me here. I never saw myself making it this far, and I sure didn't expect the battles I'd have to fight along the way. But every decision, every screwup, every hardship shaped me into who I am today. If you'd asked me 3 years ago if I thought I'd be standing here, I would've laughed in your face and said, "Not a chance." Let me tell you why.

The Boulder on My Shoulder

From 15 to 22 years old, I carried a chip on my shoulder the size of a boulder. I was hateful. Mean. I didn't give a damn about anyone's feelings. I'd walk into a room with my nose high in the air, judging people before I even knew their names.

Negativity was my armor. I'd use it to keep people at a distance and drag others down. I was angry at the world, and alcohol became my escape. Every shot I threw back, every bottle I polished off made me feel untouchable. The more I drank, the more I felt like nothing could touch me.

Honestly, it's a miracle I'm even here to tell this story. I was reckless and made decisions that should've killed me. I hurt people who actually cared about me, pushing them away before they had the chance to hurt me. I left a trail of destruction behind me without a second thought about the mess I was creating.

A Night I Shouldn't Have Survived

September 15, 2016

One of the dumbest decisions I ever made came when I was 16 years old, blackout drunk after a night of partying like I had something to prove. Five sleeves of Fireball, half a bottle of Black Velvet, a bottle of Smirnoff—probably more, if I'm honest. I drank until I couldn't feel, until standing up meant hitting the floor. I got off on being the most wasted, loudest, wildest one at the party. I made it my mission to drink enough that I'd fall on my face trying to walk. And somehow, I never ended up in the hospital getting my stomach pumped, which still blows my mind—because I drank enough to kill a horse.

That night, I made the brilliant decision to climb on my Harley and try to ride home, which was 45 minutes away. "Try" is the key word. I couldn't stay upright. Making turns? Forget it. I couldn't even hit the speed limit. I tipped over and couldn't keep the bike upright anymore. Two of my friends were following in their car. One jumped out, took my bike, and rode it back while I got in the car. It's a miracle I didn't end up in a ditch, detained behind bars, dead, or killing someone else.

Am I Really Cut Out for This?

June 6, 2022

I was still figuring out what I wanted from life. Firefighting wasn't my priority—in fact, I didn't care much about anything, and definitely not about myself. I dragged myself to the volunteer academy, physically present but mentally checked out, going through the motions like showing up was doing everyone else a favor. My motivation was nonexistent. I avoided studying, made excuses, and hoped for miracles on examination day.

Somehow I passed, but scraping by wasn't something to be proud of. It felt like a slap in the face. I clung to the title "firefighter" without putting in the work. I graduated at the bottom of my class and earned the reputation to match. I was exactly the kind of firefighter you wouldn't want next to you in an emergency. I had a choice: to stay the same and keep blaming the world or to take accountability and change.

August 28, 2023

When I stood in front of the selectboard that evening, all the bad decisions, the anger, and the mess flashed through my mind. I'd come a long way. Every poor choice, every close call, every moment I hated myself for who I'd become taught me something. It shaped me into someone who deserves to wear this badge, take this oath, and do this job.

Firefighting isn't about the title. It isn't about looking good in a uniform or holding a cool piece of equipment. It's about being ready when someone's counting on you. It's about being welcomed and trusted into the homes of strangers on the absolute worst day of their lives. It's about being a good human being who gives a damn. I had to learn the hard way before I could become a role model for others. And I had to go through hell before I was ready to become a firefighter and role model.

The Steps It Took to Get Here

May 1, 2021

I interviewed for my call department. The deputy looked me in the eye and said, "You're no longer just Nichole O'Donnell; you're Firefighter O'Donnell." That hit hard. I realized I'd taken on responsibility I hadn't fully grasped before. But I was still battling anger, alcohol addiction, and poor choices. I wasn't entirely there yet; I had one foot in, one foot out.

Between the milestones, I wrestled daily with my demons. I kept showing up, but I wasn't fully committed yet. The turning point was coming, even though I hadn't seen it yet.

August 1, 2024

I marked the start of a new chapter—one year alcohol-free. That decision to get sober was the first brick in rebuilding myself. I didn't know it yet, but the rest of the month would bring another milestone that would solidify the path I was on.

August 28, 2023

The night I was appointed by the selectboard to my career department, I texted my friend, Deputy Chief Mike Hatch: "I got the job!" His reply came quickly: "Us old folks are going to bed. Congrats and remember, the next goals are a productive career and a healthy retirement". His words stuck with me. That night, I realized the decision I made on August 1st wasn't just about quitting alcohol—it was about building a future worth staying sober for. That appointment became the turning point where everything I'd fought for finally started to come full circle.

That night tied everything together—the decision to get sober, the discipline to rebuild, and the purpose behind it all. It wasn't about when it happened, but how each step built on the last to get me here.

You Can't Fix a Problem Without Acknowledging It

November 13, 2023

I walked through the doors of the career recruit academy, to start the program I had dreamed of for so long. This was my chance to fix my past mistakes. No more just showing up. No more going through the motions. I gave it everything I had. I showed up focused. Disciplined. Driven. I went from scraping by to graduating high in my class. And none of that happened by accident; it involved a choice.

I chose strength. I chose work ethic. I chose to show up as the firefighter I knew I could be. This time, I wasn't just chasing a title. I wasn't looking for praise or glory. I was living up to what the title means. For the first time, I felt my potential come alive, and it felt right. I refused to let my past define me anymore. My bad decisions, my poor attitude, my old habits, they weren't going to stop me from becoming who I was meant to be.

The past doesn't have to dictate your future. We all mess up. We all fall short. But what defines you is what you choose to do next. Own it. Fix it. Grow from it. Don't just wear the uniform: Earn it, every single day. Work in silence; let the results do the talking, and show them who you are through action, not words. Because one day, someone younger, someone lost like you once were, will look to you for answers. And when that day comes, you'll be ready.

Key Takeaways

Your past doesn't get the final say; your next choice does. I made mistakes, carried anger, and let bad habits define me for years. But none of that had to control who I became. What mattered was choosing to change, one decision at a time.

You can't fake your way through life or the fire service. I tried scraping by, hardly studying, barely caring, hoping to get lucky. Whether in the academy or in life, that approach won't carry you when it counts. You have to put in the work.

When life knocks you down hard, that's when your true character is built. The moments that nearly broke me, the close calls, the bad choices, the nights I hated myself, those were the moments that forced me to rise.

A badge or title means nothing if you don't back it up with action. Whether you're a firefighter, a leader, or just a decent human being, what you do matters more than what you're called.

Excuses keep you stuck; accountability sets you free. I blamed everyone else for a long time. The day I looked in the mirror and owned my part was the day I started to climb out of the hole I'd dug.

Growth is supposed to be uncomfortable. It's messy. It hurts. It challenges who you think you are. But that's what shapes you into someone stronger, someone worth looking up to.

Work quietly; let your results speak loudly. You don't need to brag or seek attention. The people who matter will see your effort through what you accomplish, not what you say.

Trust is earned, not given. Whether in the fire service or in life, people are counting on you—especially in their hardest moments. Don't take that lightly.

The next generation is watching. Someday, someone younger, someone struggling, will look to you for answers. Be ready to give them something real, not just words.

No matter how many times you fall, you can always choose to rise. The fire may leave scars, and life may too, but both can forge you into something better if you let them.

Your journey proves that resilience is built in the hard moments and integrity is shown in what you do next. Keep choosing to rise.

Final Thoughts

Every decision I've made—good, bad, and everything in between—led me to exactly where I am today. And I can say without hesitation that this is the best job in the world. I'm not perfect and can't change my past, but I've learned from it. Those experiences shaped me and gave me the strength and compassion to help others in their worst moments. When someone invites you into their home on the hardest day of their life, they're trusting you to do your job well and show them kindness. Never take that for granted. If you choose this career, make sure it's for the right reasons—not for the glory, but because you want to make a difference.

Growth isn't supposed to be easy. It's uncomfortable, it's hard, but it's necessary. That's what shapes you into the ultimate underdog. I wouldn't be who I am today without the challenges I faced. Without those hard lessons, I wouldn't have the compassion I carry with me on and off duty. In this job, as in life, you're going to make mistakes, and that's how you learn and become solid.

One day, you'll be the one the younger firefighters turn to for answers. So if you screw up, own it. If you don't know something, ask. Stay curious. Make mistakes, learn from them, and get ready to come out stronger on the other side.

Reflection Prompts

- ✠ How has working in the fire service made you grow as a person? What challenges have you overcome to be successful?

- ✠ What are some areas in your life where you know you need to improve?

4

CAPTAIN COFFEE CHAOS

October 30, 2023

One week after I was hired, Joshua Sterls, one of my long-time mentors in the fire service, told me, "Never let the coffeepot run dry." But what he failed to tell me was that overfilling it is an even bigger crime, especially when it turns your captain into Gordon Ramsay but with a caffeine addiction! Typical O'Donnell trying to be the hero of the morning by making sure the coffeepot was full. But instead of filling it up like a normal person, I decided to drown the kitchen in coffee. This all happens when the captain is around—because why would I make a fool of myself in private, with no audience to appreciate my misery?

When you fill the reservoir of our coffee maker, coffee immediately starts flowing. This is why we have two pots: one to measure the amount of water and the other to catch the coffee. Well, someone had broken the other pot hours earlier, which left me with no way of measuring the right amount of water to fill the reservoir. Just having finished my ambulance check and enjoying the calm before the storm, I decided to prefill the singular coffeepot with water and then pour that water into a coffee cup, counting how many cups it filled until the pot ran out. It was tedious, but that's the method of measurement I used to fill the coffeepot exactly.

Knowing my history of failing math, though, I must have miscounted how many times the coffee cup had to be filled with water to fill the coffeepot, because I was hit in the next room with an aroma like the entire Starbucks menu. I sprinted back into the kitchen to the scene of a disaster—a coffee tsunami that had overrun the counter, the toaster, the floor, my pants, and my boots. There was coffee in every crevice imaginable.

Several rolls of paper towels later, the mess was finally under control. I looked at the coffeepot, which was somehow still full! I'm convinced the coffee maker had it out for me, possibly with a little help from my fellow firefighters, who were enjoying the entertainment.

First Run of the Day

After that catastrophe, I thought I deserved a nice hot cup of coffee. How naive! I poured a mug, took a sip, and headed back to the apparatus bay. I placed my coffee down in the ambulance, and a tone came in. I've got my radio clipped to my belt loop, and I'm packing what I call my "90 pounds of assets." In my haste to stand up, my assets decided it was time for a showdown with my coffee mug.

The mug didn't stand a chance. My butt slapped it right off the shelf. Smash! Glass and coffee catastrophe everywhere—round two in the span of a single hour (I'm the gift that keeps on giving). All over the freshly mopped floor, my pants, and the stretcher with bright white sheets. It was a nightmare. Our call was delayed, and we arrived fashionably late to the medical emergency on O'Donnell's borrowed time, with the ambulance smelling like a coffee shop. Thankfully, the patient didn't have a raging allergy to caffeine, and they even nicknamed our ambulance "The Starbucks Emergency Room Express." From that day on, I was banned from drinking coffee in the ambulance unless it was in a sippy cup.

Key Takeaways

If you're going to make a mess, make it a memorable one. When the firehouse is filled with chaos, laughter, and the occasional disaster—those are the moments that stick with you. Beyond the comedy of errors, the real takeaway is this: Mistakes will happen, especially when you're new, and you will have moments when you feel like a walking catastrophe. The important part isn't avoiding mistakes at all costs, it's how you own them, clean them up (literally and figuratively), and carry on without letting them crush your confidence.

I learned that the firehouse is a place where humility, accountability, and humor are key. Whether it's coffee flooding the kitchen or a bad call on the fireground, your crew wants to see that you can take responsibility, laugh at yourself, and get back to work. The sooner you can stop sweating the small stuff and embrace the chaos, the faster you'll grow. In the end, every clumsy moment is just another step toward earning respect and becoming part of the team. So go ahead, spill the coffee, break a mug—just be ready to mop it up with a smile.

Final Thoughts

At the end of the day, it's not the perfectly brewed coffee or the spotless kitchen that makes you a good firefighter; it's how you handle the spills, the stumbles, and the unexpected messes along the way. The firehouse is a place where mistakes become memories and blunders turn into bonding moments. So when you find yourself knee-deep in coffee (or whatever your version of chaos looks like), remember that it's not about being flawless but about being resilient, accountable, and able to laugh at yourself as you learn. In this job, the ability to clean up your messes—both on and off the fireground—matters far more than never making them at all.

Reflection Prompts

- ✠ Have you caused a coffee catastrophe or experienced a similar embarrassing situation?

- ✠ What's a lesson you learned on the job by making a mistake?

5

SELF-MADE

December 2023

Doing what scares you shows who you really are. I used to rely on someone else to make me happy. Even though looking for happiness in another person worked for a while, it didn't last. When I had a significant other to make my life happy, I wasn't doing things for myself. Instead I was constantly trying to please everyone around me, worrying about what others thought, always looking over my shoulder, and trying to control things that were completely out of my hands. I was stuck in a never-ending spiral, using someone else to fill a void that only I could fill.

Walking away from the longest relationship I'd ever been in midway through the academy weighed heavily on me. My focus, my studies, everything was on the line. The decision was now or never, and it felt like it would make or break me.

One day something shifted. I knew I needed to change, and no one else could do that for me. Until then, when it came to working out, I would go only when my partner went. I wouldn't do things on my own because I was trapped in a cycle of codependency, which was tearing me apart. If I wanted to succeed as a firefighter, I had to make the change. Leaving that long-term relationship was the best decision I ever made for myself. This job means making sacrifices, so I didn't care what it took; I knew that it was exactly what I needed.

The fire academy is intense enough without having to balance a long-term relationship. Imagine being completely submerged in your studies, laser-focused on passing so that you can finally start your dream career as a full-time firefighter. Distractions are the last thing you need when you want to focus on your career.

You're probably wondering why I would leave a relationship for a job. But this isn't just a job to me; it's a lifelong career, a commitment, and a choice I've made for myself. It may sound selfish, but I couldn't do otherwise. With 5 weeks left in the academy, we were deep into the most grueling part of our training. We were 50% done, and I was stressed to my core. Do you think I wanted to deal with anything else? Nope. But it was gnawing at me: I couldn't keep pursuing this relationship beyond graduation; I knew what I wanted, and it wasn't this anymore.

I'd become so focused on my career that nothing else mattered. I felt trapped, like I didn't even know who I was anymore. Something had to change, and that voice inside my head kept telling me it was time to move past the relationship.

So I made the call and ended the longest relationship I'd ever been in after 3 years. I was scared, exhausted, and numb—completely shell-shocked. I kept replaying the what-ifs in my head and inevitably returned to the same conclusion: I needed to figure out who I was, and for once, I had to put myself first.

It was the best decision I ever made. The day I walked away, a huge weight was lifted off my shoulders. I became more independent than I'd ever been. I taught myself how to cook, how to write, how to take care of myself, and how to do things just for me.

April 2025

Life has a quiet way of circling back. Almost 2 years after I chose to walk away from a decision that shaped so much of my personal growth, I got a message from my ex. He said he had found something of mine and wanted to return it. The conversation was surprisingly kind. He was more open, more thoughtful than I remembered him being when we were together. It felt different, like I was talking to a familiar stranger. I agreed to meet, not because I was looking for anything, but simply to close that loop. When I saw him, there wasn't any big emotional reaction—just clarity confirming that this chapter really is closed and gently reminding me why I left in the first place. I didn't feel the need to revisit the past or hold onto anything. I saw it for what it was: a part of my life that taught me a lot and helped me to grow into the person I've become.

What did I learn? That relationship taught me how to listen to myself, how to recognize when something no longer fits, and how to move forward—without needing to look back or seek reassurance. There's no resentment, just calm appreciation for the lessons it taught me and the role it played in getting me here. Some moments in life don't stir up the past, but quietly affirm that you've grown beyond it. That was one of those moments. Not everything needs a dramatic ending. Sometimes, the softest closure is the most powerful—leaving you with peace, not questions, and gratitude, not pain. That chapter served its purpose, and like a fire we've already knocked down and walked away from, there's no need to go back and stir the ashes.

Key Takeaways

Courage means doing what scares you. True growth happens when you face your fears and make hard decisions that are best for you, even if others don't understand.

Happiness can't come from someone else. Relying on a relationship for your sense of worth or joy leaves you feeling empty when it ends. Real happiness starts within.

Independence is earned. Leaving behind what no longer fits, even when it's painful, allows you to discover who you truly are.

Closure doesn't need to be dramatic. Sometimes the quiet moments bring the clearest peace and affirm how far you've come.

Your strength builds trust. When you learn to rely on yourself, you become someone others can rely on too, especially in a demanding career like the fire service.

Your only true responsibility is yourself. It's not about changing others' opinions. It's about staying true to your path and striving to be better each day.

Final Thoughts

In the end, being self-made means realizing that you are the one who holds the key to your own happiness, peace, and success. Stop seeking validation from others, make decisions that honor your well-being, and build the life you want. By letting go of what no longer serves you and standing on your own, you can experience true growth—to become the strongest, most authentic version of yourself, the kind of person you can count on, and the kind of person others can trust in the toughest moments.

Today, I am the woman who would have protected me as a child. I cook, I clean, I take care of myself, and I've found happiness in myself. I live on my own in a beautiful carriage house, and I am at peace finally. Finding joy in the smallest things, I no longer care what anyone else thinks of me. I have come to terms with myself, being happy on my own, and never having to rely on anyone ever again—just like this line of work. If I were always to rely on someone else, then how could I ever think my crew would be able to rely on me? They wouldn't be able to if I wasn't independent enough.

At the end of the day, everyone has their own opinions, and you're not going to change their minds. Accept this and keep moving forward; you'll be so much happier. Yes, it's easier said than done, but it's possible when you put your mind to it.

I thrive in the version of life I want to live by no longer trying to live someone else's life. The only person who's going to take care of you for the rest of your life is you. So be kind to yourself, keep a positive mindset, act with good intentions, and don't be jealous. Focus on being the better today than you were yesterday. Everything comes with time.

Reflection Prompts

- What's something you've had to sacrifice for your career in the fire service?

- Are you comfortable with being on your own? Why or why not?

6

CHALLENGES YOU'LL FACE

Early December 2023

Practical evolution: hands-on search-and-rescue scenario at the fire academy. This evolution simulates finding potential victims in a house with nonideal conditions. Paired up, assigned as the lead, I'm doing a left-hand search with my partner right behind me and the instructor at my hoist, observing.

Everything was going smoothly, until suddenly I heard and felt that all too familiar, gut-wrenching pop and twist in my shoulder. A sharp, searing pain shot through me like a bolt of electricity. My body froze, and almost immediately I knew what had happened: shoulder dislocation. Dislocating your shoulder midsearch is a firefighter's worst nightmare. You can't just pause and call a time-out. The practical examination was still going on, and my team were relying on me to finish. Without thinking, I yelled out a curse that echoed through the structure like a siren, loud enough for everyone to know something was wrong.

The instructor asked, "What's going on?"

"My shoulder just popped out of the socket!"

Not exactly what you want to admit in the middle of a search-and-rescue exercise. Everything felt like it was in slow motion. My mind raced between thoughts of *How do I fix this?* and *I can't let this injury take me out of the academy.*

The instructor sent me to EMS, but before they even had a chance to assess me, I grabbed my shoulder, clenched my teeth, and ripped it back into place. The feeling of my shoulder popping back was as excruciating as it was relieving—and any relief was short-lived. A sharp, burning pain followed to remind me that my shoulder was shot.

Pushing Past Your Limits

I did what any stubborn firefighter would do: I played it off like nothing was wrong.

"I'm fine," I claimed, despite the aggravating pain. I moved my arm around to show them at EMS that I still had "full mobility"—even though each movement felt like my arm was being ripped off. I dropped down and cranked out 10 push-ups even though every push felt like fire through my arm. In my mind, quitting wasn't an option. I preferred to suffer temporarily rather than lose my dream job or repeat the academy.

Late December 2023

With less than a month left of the academy, I'm in the middle of a 24-foot ladder, performing a single-person emergency throw, with Mr. Hatch next to me, grading my skill. My bad shoulder was wrapped tight in kinesiology tape—the only thing holding it together.

Midthrow, I felt the all-too-familiar pop and twist, and I knew my shoulder was done for. The pain shot through me like a blowtorch. I slammed the ladder to the ground, took a breath, and voiced, "Yup, give me a minute." Mr. Hatch gave me a moment's break to reset my shoulder, yet again. I had become so accustomed to the pain—which was intense—that resetting it was just another part of my day. I wasn't going to let it sideline me with only 2 weeks left. There was no way I was starting over.

The Final Stretch

January 20, 2024

The Massachusetts Firefighting Academy has three campuses: At Springfield, where I went, they call the probies dragons; Stow is enormous, the mother lode of new recruits for the state; and Bridgewater is brand new.

Our hazmat experience fell 7 days before graduation. This was one of the coolest experiences of my life. You could watch instructors blow stuff up in a controlled environment, and you could play with propane, liquefied natural gas, pressurized vessels, grill fires, and things you don't typically see in the field.

We took part in this experience at the Stow campus, an hour and a half away, since we don't have the resources available to do this at Springfield. A lot of the experience was observational—witnessing and not so much touching. Still, you need to be aware of what it was like to be in a hazardous situation to meet the requirements of the program, so that you aren't thrown to the wolves when you begin the job.

In short, you have the opportunity to experience how dangerous hazmat is without getting hurt. Unless you're Probie O'Donnell—then you always find a way to become a safety topic of what not to do at the next union meeting.

The Hazardous Gas Practical Experience

It was brutally cold, icy, and frigid all day long, and I was unable to get warm even in 100 pounds of turnout gear. Hose was laid everywhere, frozen to the ground, with nozzles cracked open to prevent the hoselines from freezing with water trickling turning to ice the minute it's exposed to the air. Once again, instructors yelled in all different directions; this time, however, we calmed the chaos.

Any evolution was more about getting to experience what it's like being near an extremely dangerous situation. For example, for a fire in a pressurized vessel holding propane, we were lined up in groups of five or six recruits on the hoseline, with an instructor at our hoist. They ordered us with step commands to move forward on the hoseline to the vessel—hissing because it was so hot and ready to rupture—and then told us to open up the nozzle and directed us on where to aim (not in direct flame, etc.).

I began my day walking alongside the metal grates on the fireground at the Stow campus and suddenly lost my footing, crashing down with an abrupt stop on the frozen pavement. Apparently metal grates are slick when they freeze—why wouldn't I know that? (Face palm. Because it's commonsense.) My hands instinctively shot out to break my fall, and I sprained my wrist.

I took some ibuprofen and just dealt with it. Falling first thing in the morning made the day agony from start to finish. Not only was I hurt, but the cold made it 10 times worse. Less-than-ideal temperatures amplify whatever pain you have to begin with.

This hazmat day was a requirement, and if you didn't partake in it, then you would be kicked out. I couldn't report an injury, or I would be forced to do the academy all over again. I faked being fine until I made it in my assigned group—going the whole day showing no signs of pain. My wrist was useless, my shoulders were shot, and I was both physically and mentally exhausted.

I had already pushed through two shoulder dislocations in the previous weeks, so there was no way I was letting a sprained wrist stop me from completing the program. This tested my mental resilience more than anything. I wouldn't recommend working with three painful joints, but when your job depends on getting through it, you do what you must to survive.

Key Takeaways

Failure doesn't mean you suck; it means you're learning. If there's one thing you should take from my academy journey, it's that failure is part of the process. It's your first attempt at learning. Don't let it define you; instead, let it drive you to improve.

Injuries and setbacks will happen, and how you handle them shows your grit. I dislocated my shoulder twice, sprained my wrist, and still kept going. The academy (and this job) isn't about avoiding hard times; it's about proving to yourself that you can push through them.

Pain is temporary; accomplishment is forever. Every injury hurts like hell. But the pride I felt crossing that graduation stage? That stays with you. The hard days are what make the moment of accomplishment worth it.

No one's coming to save you midsearch. Own your struggle. When I got hurt during search and rescue, there was no time-out button. In this job, you have to think fast, act smart, and do what needs to be done without waiting for someone else to step in.

You're stronger than you think, mentally and physically. The academy isn't just about muscles. It's about your mindset.

You'll find yourself in moments when it feels easier to quit than go on, but don't quit. Resilience is built when you just grit your teeth and get through it. Would I recommend resetting your own shoulder or faking it through a sprain? No. But when the job is on the line, you do what you must.

Show up, even when it's hard. That's the mark of a firefighter. You took an oath. That means pushing through the suck, being accountable, and showing your team that you can be counted on even when you're hurting.

If you put more good into the world than bad, then you'll be just fine. At the end of the day. this career is about serving others. Let that fuel you through every challenge you face.

Final Thoughts

The academy isn't just about surviving physical challenges; it's about proving you have the grit to push through, no matter the challenge. Each injury was a test of resilience. Remember that pain is temporary while the accomplishment when you apply yourself is permanent.

These injuries weren't just setbacks; they were moments that tested my resolve. I could have quit, but I didn't. Once you pass the academy, this job tests your limitations. Having taken an oath to protect and serve, we hold ourselves accountable for the rigors and physical demands of the job. Pushing through resilience and coming out stronger on the other side—that's what this job is all about.

Reflection Prompts

- What were some challenges you overcame during your academy experience? How did you feel afterward?

- What would you have done differently at the academy, and why?

7

BLINK AND IT'S OVER

January 27, 2024

As a recruit, I was perceived to be "too happy to be here." If I had a nickel for every time someone told me I looked too happy, I'd be rich. One of the support staff helping out the instructors even nicknamed me Smiley. I'm commonly asked, "Why are you always smiling? What's your secret?" My standard reply is, "If you walk in here every day with a poor attitude, it impacts others negatively. Be the light that shines so bright that when others try to look away, they can't."

No one wants to be at the academy. You're working Monday through Friday, not two 24-hour shifts a week. All it takes is one hiccup—whether an injury or a failing grade—and then your job goes bye-bye or is put on hold. For some, this job is a calling, and for others, it's a dream. Starting each day with a positive mindset makes all the difference. Simply put, a positive mindset leads to a positive day; a negative mindset leads to a negative day.

Welcome to New England Firefighting

Most days at the recruit academy were brutal, counting down the hours until you could crawl into bed and praying that your body doesn't ache for the next 3 days. I don't know what was worst: the workouts; the temperatures (from cold to freezing) in the drill yard; getting yelled at by the instructors; or all of it combined.

Burn days were my favorite. We had six of them. Each one is basically a level up in *Firefighting for Dummies*. The first couple of burn days were straightforward—getting the nozzle to the first or second floor and putting out the fire. By the second week, though, it seemed like they were trying to kill us. Ladders were everywhere, like a firefighter jungle gym. The snow was coming down so heavily that it felt like we were training for a winter survival expedition, and sheer ice made a figure-skating rink out of the drill yard. The recruits who were able to stay on their feet deserved gold medals.

Winter in New England is something I wish all firefighters in the warmer regions of the United States could experience. Hot weather is one thing; wildland is another. But frozen conditions are a different breed. Your gloves stick to just about anything, you have to crack open the nozzle on the fireground so the line doesn't freeze or burst, and you struggle to stay warm in your turnout gear even while battling the hottest fires.

Embracing Chaos

Let me tell you about my *not* finest moment during an evolution. There I am, hauling this ridiculously heavy 2½-inch hose up to the second-floor mezzanine. My two partners are behind me, frantically trying to shove more hose up the stairs. I'm leading the charge, feeling good, thinking I've got this. Then I hit the doorway and wham! Suddenly, gravity turned against me.

My feet, my hands, my whole body just gave up. Next thing I know, I'm sprawled out on the ground like a chalk outline at a crime scene. I landed face-first, right on the pistol grip of the nozzle. I had zero control over my body and simply laid there, accepting defeat: "Well, this is it. This is where I die."

The instructor looked down at me and says, "Shoot, you all right, Nineteen?" (that was my recruit number). All I could muster was a half-dead grunt, "Yeah"—not convincing anyone, least of all myself. But with work to do, I got up, pain and all, and kept going. I dragged that damn hose to the third floor, found a victim along the way, radioing "Victim, victim, victim;" then found the door behind me, radioing to command, "Victim is out the Charlie-side door, second floor, we need EMS now;" and then knocked out the fire. Because that's what we do.

Firefighting might hurt, but we suck it up and get the job done. We learn quickly that this job is both physical and mental. Yeah, I was hurt, but I was equally embarrassed that it happened; still, I convinced myself to finish the evolution. This is the lifestyle of a firefighter, and there will be more days like this and even more obstacles to come. Understanding how to mentally navigate and strengthen yourself through adversity is what separates the people who really want this job from those seeking glory.

Frozen Gear

Then there was that frigid New England day when it was so cold my gloves froze to the railing of the mezzanine. This is essentially a second-floor catwalk used by instructors to observe drills from above. (If you screw up, someone's definitely watching.) During an evolution, I'm walking up the stairs, my gloves touch the metal railing, and suddenly I'm forced to an abrupt stop, making me useless to my teammates. It took every ounce of strength I had to rip my gloves free. It's a miracle I didn't shred my gloves into tiny pieces.

Don't get me started on the air packs. I had two instructors yanking on the straps on both sides of my air pack, trying to loosen it. My waist straps were so frozen that they were basically welded shut. Two grown men pulled on my waist straps just to get the pack loose enough for me to wiggle out of it. This is New England firefighting, so welcome to the reality of the job!

Another time, we were handling fully extended 35-foot ladders. These are roughly 135 pounds but manageable if you train on them. A requirement at the recruit school is to have three people to carry a 35-foot ladder at all times—even though on the fireground, you might sometimes have only yourself. Lucky me, I was the tip person, at the front of the ladder, while the ground is like an ice rink. As we begin lowering the ladder, I lose my footing. My boots might as well have been lubricated because I'm sliding all over the place like Bambi on ice.

Before I know it, three instructors come sliding in like penguins on a rescue mission. One grabs the ladder, the other two grab me by my air pack to stop me from crashing to the ground. Too late! I'm headed into a full split—and I am not at all flexible. Next thing, I'm feeling muscles stretch in ways they were never meant to. Accepting my fate, I lay there helpless, gaining one hell of a story and a new-found respect for yoga. My instructors had saved my noggin from getting smashed by a large ladder, but my muscles laid themselves to rest for the rest of that day.

Our final burn day was the cherry on top of the disaster sundae. We finally got radios so that we could pretend we were on the fireground. Of course, that came with more responsibilities, and they threw us into Mayday situations—particularly the kind where they snatch one of your crew members like it's *Naked and Afraid: Fire Academy Edition.* Suddenly, our three-man crew is down one member, and it's up to the two of us left to figure out that we now have a missing firefighter.

If you were too ignorant to realize that one of your crew members was missing, they'd hear the chewing out you received from the instructor leading your evolution in the next county over. We were rescuing victims, working as the rapid intervention team, operating the aerial, ventilating the roof, and hauling the 2½-inch hose up three stories fully charged. Water weighs 8.34 pounds per gallon, and 2½-inch hose flows 250 gallons per minute on average. Now do the math: 250 × 8 pounds gives you 2,000 pounds (at minimum). If day one was a warm-up jog, by day six, we were in the firefighter Olympics without the cheering crowd—just instructors yelling at you and watching to see if you'd eat it.

Lessons Learned

Ridiculous moments like the time I did the splits on ice seem comical now, but they remind me of a deeper message: It isn't just about surviving the training; it's about learning to trust the people around you. And trust doesn't come easy.

No matter how tough or prepared you think you are, there will be times when you need someone to catch you. That day, it was instructors pulling me back from an injury. But on the fireground, it's about trusting your team to have your back when it really counts.

That trust didn't come naturally with my academy group, unfortunately. As a new recruit, you interact with people with many different personalities, including a range in enthusiasm for being there; some may seem to be merely present, while others are overzealous about being at the academy. The true lesson of the academy is to know why you're there and get everything out of the experience, so that when you're done you can apply that knowledge as your foundation of your first year and ultimately your career.

I wasn't anyone's first choice when it came to partners. Every workout I'd be paired with an instructor instead of a classmate, which made me feel like I wasn't really "one of them." This prompted a flashback to when I used to get bullied as a kid.

One morning as I walked in, a recruit commented, "Did you do a line of coke this morning? You're too happy and have way too much energy." He looked at me like I was nothing, like I didn't belong there. Another classmate barely acknowledged my presence, talked over me, or acted like I was invisible. Another threw the classic line, "I've been on the job 7 years," at me like he was better even though we were both in the same class.

While some of them were focused on where they'd go to drink that night, I was there to learn, to get better, to give everything I had. Not everyone was like this, but I'll never forget how those people made me feel. Each week, I counted down the days until graduation, coaching myself, *I got myself here. Once these 10 weeks are done, I never have to go through this crap again. Then the real job begins.* Not everyone in my class was intolerable, though, and I did make a few solid friends who had my back and made the grueling days a little more bearable.

The lesson I took from this is that you're not always going to like everyone you meet or work with, but that doesn't mean they're bad at their job. I remember an instructor prompting, "Look around. What do you see?" I glanced around the room, filled with a sea of recruits in turnout gear each carrying their own stories, struggles, and reasons for being there. He continued, "No matter what we believe or how we live, we're all here for the same reason. We wear this badge, we carry our last names on the back of our coats, and that's what matters." That hit home because it's easy to judge someone or write them off without getting to know them first, but in the end what matters is showing up on the fireground prepared and ready to work, because feelings don't exist there.

Key Takeaways

Attitude sets the tone. Your mindset walking into the fire academy (or into any challenges) shapes not only your day but also the energy you bring to those around you. Positivity is contagious.

Adversity builds resilience. Whether you're slipping on ice, battling frozen gear, or pushing through exhaustion, the tough moments forge your mental and physical strength.

Trust is earned in chaos. In this job, you'll fall. What matters is knowing someone will be there to help you up—and being that person for others.

You can't choose your crew, but you can choose to work as one. Not everyone will be your friend, but when it's time, what matters is having each other's back on the fireground.

Focus on why you're there. Block out the noise—from doubters, egos, or distractions—and stay committed to learning and growing.

Pain is temporary, but the impact is lasting. Every bruise, fall, and hard day prepares you to make a real difference when it counts.

Final Thoughts

Even though my time in the academy wasn't exactly glorious, those moments of pain, chaos, and humiliation forge us into a team. Falling on your rear in front of everyone, picking yourself back up, pushing through exhaustion and bruised egos—those are the experiences that build something real connecting you and the people around you. Whether you're a volunteer or full-timer, when duty calls, the fire doesn't care about your doubts or your exhaustion. The only thing that matters is getting the job done.

Just like you can't pick your family, you can't always choose the people you work with—but you find a way to make it work. Put your ego to the side and remember what you're there for in the first place. Pain is temporary, but the lives you save go on. When that academy training kicks in during real situations and you see the difference you're making, that's when it all clicks, and you understand exactly why you're there in the first place.

Reflection Prompts

- ✠ What's an embarrassing experience you had at the academy? What did you learn from it?

- ✠ Was there a sense of camaraderie in your graduating class? Are you still friends with any of your classmates?

8

The Salty Rookie Mentality

Welcome to the firehouse. Here's what they didn't teach you in that classroom: how to be a human on the job. They covered hydrant operations, ladder throws, grunt work, and fire behavior, but no one pulled you aside to explain the mindset it takes to survive this job. The attitude. The respect. The awareness. The little things that make you the person someone wants next to them at 3 a.m. on the worst day of their life.

This chapter is the class you never got in the academy. No PowerPoints. No textbook. Just rules from real life, learned through hard lessons. I've picked up a set of rules through mistakes and watching those who do it right. They're not official standard operating procedures, but they're just as important. Because at the end of the day, being a good firefighter means more than just knowing your job; it means being someone who shows up with a good attitude, a willingness to learn, and the kind of presence that lifts the team up instead of weighing it down.

Every firehouse has its own culture. What works in one department might not in another, and that's what makes this job so great. There's no one "right" way to do things, just whatever works for your response area. What matters is the mindset and discipline you bring. You can do a thousand things right, but one bad attitude or lazy habit can stick with you forever.

Build a strong foundation—not just in skills, but in character. This is the stuff I wish someone had told me.

The advice in this chapter was adapted from bullet points that I originally shared on social media, along with pictures taken at the firehouse. My first post got almost 100,000 views, and the next post surpassed my expectations, getting 120,000 views and over 700 shares. The biggest names in the fire service shared my posts, including Chief Miller. I realized I had found my niche.

After that, I began to notice how many like-minded people there were in the fire service. This is when the underdog began to shine: I began to network with those who I never thought would notice me. I had an opportunity and ran with it. That's one of my biggest pieces of advice: Take every single opportunity you're given.

The Salty Rookie Mentality began as an intuitive experiment and turned into my niche in the fire service culture, allowing collaborations with big names in this field. The rules that follow have determined my mindset and how I operate:

1. Be the firefighter you want showing up to your house to save your family or you.

2. Ninety-five percent of success is showing up on time and not having a bad attitude.

3. If you're on time, you're late; if you're early, you're on time.

4. Listen to all the gossip, but *never* spread it. If you hear something you weren't supposed to, then no, you didn't hear it. If someone asks for your opinion on someone or tries getting you to feed into the gossip, the best reply is walking away. Do not be the messenger. Stay in your own lane. Those rumors you hear can be career ending. Don't get involved. Once you insert yourself into the conversation, it may turn around on you, and you may find yourself being blamed as the catalyst of the drama.

5. Passion will take you places a classroom never could.

6. Know the difference between time and experience on the job—for example, 20 years in a slow environment versus 5 years in a fast-paced one. Time doesn't mean anything if you don't have the experience.

7. Don't rush to lead just because you qualify.

8. Give a damn.

9. They won't expect you to know it, but they'll expect you to care. Everyone dislikes a know-it-all; that attitude leaves a bad taste in everyone's mouth.

10. Show them; don't tell them.

11. If you want it, go for it. You never know unless you try.

12. You can't control the outcome, but you can control how you react to it.
13. Lead from the jump seat.
14. Practice like you play: Know the basics of the job and surpass expectations. Basics should be second nature.
15. Don't worry about screwups. We all make mistakes, even the most experienced among us. Ask your officers about mistakes they've made in their career, and your mind will be blown by some of the obstacles they've faced. Learn from it when you screw up so that you do better the next time. Turn your mistakes into fuel, and use that to your advantage, to succeed.
16. Don't look for validation from others.
17. Expect to be picked on, and toughen your skin. If they aren't picking on you, then they probably don't like you. If they tease you, maybe it isn't a bad thing—as long as it stops short of hazing.
18. Don't burn bridges. Networking is endless.
19. Everyone goes home, including the homeowners. Put the victims first. You have gear to protect you to an extent, whereas they have skin that's rated for 100°F.
20. Rock bottom doesn't break you; it builds your foundation.
21. Some practice until they can get it right. Practice until you can't get it wrong.

22. You don't need a perfect past to create a powerful future.

23. If it's dirty, clean it. If it's broken, fix it. If you don't know how, ask.

24. If they tell you to sweep, just sweep. You still make the same amount—unless you're a doctor or lawyer, in which case you're screwed.

25. Do the right thing when no one's watching; even when you think they aren't, they are.

26. Downtime is practice time, not in the heat of the fire.

27. The only stupid question is the one you don't ask.

28. Human Resources is *not* your friend. Never say anything to anyone you're not fine with everyone else knowing. Deal with it behind closed doors, unless someone touches you inappropriately, in which case, you must go to Human Resources for safety's sake, for yourself and others in the department.

29. You can lie to the world but not to the person in the mirror.

30. If you don't half-ass it, you'll never have to question if you did your job well enough.

31. Never let your officer or senior member carry something. If they go to pick something up, either offer to take it for them or just do it. Do not let them do

anything other than their own job on the fireground. Don't overstep, but don't sit there empty-handed.

32. Social media can ruin you. Be careful what you post. Don't post anything about your job your first year. Keep your opinions to yourself. Don't express beliefs or politics. If you have to ask yourself if you should post it, then that means you shouldn't. What you do in your personal life is no one's business.

33. If you need help, look up the answers yourself. Then, if you can't find the answer you were looking for, ask. If it's safety related, then it's OK to ask. Otherwise, try to figure it out it yourself first, before you give your officer a headache; they've already got more than enough on their plate without a probie asking for help with something that should be easily solvable (but they were too lazy to actually try doing it on their own). No one's ever going to discredit you for trying.

34. Try before you pry. It's probably unlocked.

35. Take the picture.

36. Don't bang your coworkers. Don't foul the place where you eat. If you do, you can kiss your respect goodbye and be labeled as someone who screwed their way up the ladder rather than working for it. If you want a relationship with someone in this line of work, let it be outside of your own firehouse. Be smart.

37. You're welcomed into individuals' homes on the worst day of their lives. They trust you to be there. No one else has that privilege.

38. If they aren't paying your bills, screwing you, or taking care of you, then their opinion doesn't matter.

39. Be a firefighter, not a woman in the fire service.

40. Learn to make five or six really good meals. Even if you never cooked before you started this job, you *should* learn. Full bellies make happy firefighters. If you're a good cook, your crew will respect you a lot more. No one wants a probie who burns every meal. If you buy groceries and are the firehouse chef, everyone should pitch in equally. If there's six of you on shift and the total is $100, then collect $17 from each of your teammates; don't be that cheapskate who charges everyone $20 to get a free meal. Also, don't forget dessert.

41. You are the last one to make a plate and eat. Do the dishes. Housekeeping goes a long way. You are the first one up, get a pot of coffee brewing, and have the newspaper on the table ready to read first thing in the morning. Never let the coffee pot get low.

42. Your first week working with each shift, bring food. When I began, I provided bagels with homemade cream cheese, donuts, and Danishes. You don't have to be fancy, but put forth effort. Don't show up with

nothing when you start your shifts! Everyone's in a better mood when they're fed.

43. Your ego is your biggest enemy. Learn to control it.

44. Find a fitness regimen and stick with it.

45. When you relieve the last crew, check in with the shift captain and ask how the prior shift was. Then check your gear to make sure you're ready to go and your radio battery is fully charged.

46. Your opinion doesn't matter as a probie. Shut your mouth. You may have teammates who have been doing this job longer than you've been alive.

47. Have fun! This job is what you make of it. It should be the best job in the world! It can be either a lifelong journey of endless opportunities filled with happiness or a miserable experience counting down the days to retirement, depending on your mindset.

48. Be yourself. Be personable. Be accountable. Show up when it matters.

49. Treat others the way you want to be treated—in other words, follow the golden rule. No one cares how high your grades were, but they will care how you treated them. They may never remember your name, but they'll always remember how you made them feel. You can be the best firefighter around, but if you don't have people skills, you're useless to your teammates and the community. No one wants

to work with a jerk, even one who is phenomenal at their job.

50. Work ethic beats book smarts every time. Firefighters are built on the fireground, not in a classroom. Certificates and college degrees don't mean a thing if your work ethic doesn't match them.
51. You're going to face calls that will break you, so find an outlet for your emotions.
52. Go in like God sent you.
53. One hundred percent of missed opportunities are the shots you didn't take.
54. Check in on your people. Smiles are easy to use as a cover-up.
55. Understand that if someone gets hurt or doesn't go home because you didn't care enough, it's your fault. You will have to live with that.
56. Rank doesn't make you a leader; respect does.
57. Stay off Hit'n It Hard From The Yard.
58. Do better and hold the standard.
59. Know your building construction types (table 8–1). For instance, in balloon frame, there are no fire-stops, the windows are aligned with each other, and usually there is a chimney in the center, so fire can spread up to the top rapidly.

TABLE 8–1. Building construction chart

	Construction Types				
	Type 1: Fire Resistive	Type 2: Non-combustible	Type 3: Ordinary	Type 4: Heavy Timber	Type 5: Wood Frame
Examples	High-rises, hospitals	Warehouses, commercial buildings	Downtown buildings, older apartments	Old mills, churches, historic buildings	Houses, townhomes, newer construction
Materials	Concrete, protected steel	Steel frame, metal roof	Brick walls, wood interior	Big wood beams	Lightweight wood—in everything
Features	Won't burn easily but traps heat; hard to ventilate	Won't burn but collapses quickly when heated	Fire hides in walls and ceilings (voids)	Thick and slow to burn; holds up well; gives warning before collapse	Fastest to burn and collapse
Cautions	Heat builds fast; flashover risk inside	Steel bends and fails fast—watch the roof	Hidden fire spread—pull ceilings early	Heavy fire load; rekindles common	Trusses fail in under 5 min—so get in and get out!

Key Takeaways

This job demands not only strength but also awareness, discipline, and the guts to think under pressure. Building construction isn't just textbook filler; it's life-or-death information. Knowing how buildings behave isn't about being a good firefighter; it's about staying alive long enough to become a great one. If you don't know what you're walking into, you might not walk out. The five types of structures have their own habits, weak spots, and warning signs, and it's your responsibility to learn them—not for the sake of a test, but for the crew next to you.

When the heat's rolling, the smoke's banking, and decisions need to be made in seconds, your mindset matters most. Be the one who gives a damn, who doesn't guess, who trains like it's real, and who shows up like it counts because it does.

If you want to last in this job and become worth following, then remember these rules:

- It's not about being a hero; it's about being solid. Consistent. Reliable. Built for the job.
- It starts with giving a damn.

Finally, I want to share the advice of my instructor, Mr. Boucher—take it or leave it:

- If you don't use the skills, you lose the skills.
- Slow yourself down.
- Practice like you play.

Final Thoughts

This chapter wasn't about textbook answers or examination preparation, but rather about what they don't teach you at the academy or in the firehouse, the unspoken rules learned only from experience, observation, and getting your hands dirty.

These lessons weren't just handed to me, I earned every-one one through fire, sweat, and a fair share of mistakes. They are built upon long days, hard calls, mistakes I owned, and moments that reminded me why I chose this job in the first place. You won't get it all right your first year. You're not supposed to. What matters is that you show up ready to learn, ready to grow, and ready to earn your place—not just on the roster, but in the crew's trust.

Carry yourself with humility, keep your ego in check, and treat this profession with the respect it deserves. At the end of the day, your reputation won't come from how fast you masked up or how many drills you crushed; it'll come from how you make others feel, how you handle yourself under pressure, and the kind of teammate you choose to be. Take these rules, apply what fits, and when your time comes to teach the next probie, pass them on and remember where you started.

From one probie to another, here is my solemn word: You may not always have someone to lean on, but you've got me. Learn the lessons, carry them with you, and know that you're never alone. We're in this together, and it's an honor to pass this knowledge along to you. May it guide you the way it guided me. I wish you the best in your first year, which is only the beginning.

Reflection Prompts

- ✠ What rules would you add, and why?

- ✠ Was there a mentor or senior member during your time at the academy who had an impact on your life or career? What advice did they give that still sticks with you?

9

AIR ISN'T FREE

May 2021

I never believed I'd be stepping into a full-time fire service career in only 2 years. Before the academy, I was all dreams and determination, soaking in every drill like it was the sun.

One of my most intense and humbling experiences in the fire academy was the air consumption drill. For recruits, this was the bread and butter of our training. It wasn't just about how long you could last on a bottle; it was a physical and mental gut check.

For competitive probies like me, the air consumption drill became a test of grit. Could you outlast your classmates? Could you prove you had what it takes? I wasn't great at it in the beginning, burning through my bottle fast and feeling the sting of failure every time I tapped out early. But it taught me something bigger than how to pace my breaths; it taught me that air is life and isn't free.

Whenever we go on air, we step into an environment designed to kill us. SCBA is more than gear; it's a lifeline. If you run out of air before you're out of the immediately dangerous to life and health setting, things can and will go bad fast. I didn't fully grasp that until I heard the following story, which changed how I view every breath I take—whether in training or on the fireground.

Low-Air Mayday

A friend of mine came within inches of losing his life on the fireground years ago. He had 20 years on the job—the last guy you'd expect to find in a Mayday situation. But nobody's invincible. The fire went from routine to chaos in seconds. Heat was rising, visibility vanished, and before he knew it, he was cut off from his team—trapped and lost. Those few minutes of disorientation felt like an eternity to him. His heart raced, his breaths were coming faster and faster, and all he could think about was whether he'd make it out alive. Panic took over, and he began to run out of air, burning through it like there was no tomorrow. Inevitably, his low-air alarm went off.

Just when it seemed like it was all over, he found the hoseline and pulled himself out seconds before the room flashed over. He nearly didn't make it home that day. If he hadn't found that line when he did, his story wouldn't have the happy ending it does.

His story shook me to my core. It wasn't some far-off tragedy; it was real and happened to someone I knew. It made me realize just how important air management was. One breath is the difference between life and death.

Fear drove me to make a hard reality check, asking myself, *Am I ready for that? Would I have made it in his shoes?* In response, I realized that I couldn't just show up anymore; I had to be ready for the worst every shift and consequently had to train harder, push myself further, and do better. There's no room for complacency. Every second matters; every breath counts.

Bottle Breathers

Imagine ending up in a similar situation with way less air than my friend had. Because this is a self-directed job (and one you signed up for), you've got to be ready to pull your weight without excuses. To be trusted in the thick of the action, at the seat of the fire, you need to know how to manage your air. If you can't master it, you can kiss any chance of being on the interior goodbye, and expect to be stuck doing exterior grunt work instead.

No matter how disciplined you are, there will be times when you're giving it everything you've got on a tough fire and you hear the low-air alarm. Burning through a bottle is not always avoidable, so paying attention to your air supply is essential.

To bottle breathers, the critical difference between working hard and running out of air is when you fail to put in the effort beforehand, owing to laziness. If you're gassed after pulling a hoseline and haven't given it your all, then you haven't trained hard enough, and that's on you. There's no excuse. This is a physically demanding job, so you must give a damn and put in the effort.

June 2022

Man, I suck at this! I wasn't exactly killing it with air management at the call/volunteer academy. While I wasn't the worst, I wasn't anywhere near the best, and that drove me insane. I'd get winded almost immediately, chest heaving and mind spiraling. I was mentally and physically checked out, and everyone could see it. Failing your crew is always humiliating. I wasn't the one to count on for interior work because I couldn't be trusted to make my air last—and I wouldn't have trusted me either.

We use different bottles in the fire service: Some are rated for 30 minutes, others are for 45. It varies by the department, but here in Massachusetts, the academy runs 45-minute bottles. Even with that, I couldn't get mine to last more than 19 minutes. The moment that low-air alarm sounded, my mind went into overdrive—fear hit like a freight train. This is it, I'm out. I'm done. You can feel the air slipping away, and panic takes over. You're not thinking about the fire any more or the task at hand because time is no longer on your side.

Low-Air Alarm Anxiety

I'm going to explain what goes through the average firefighter's mind when the low-air alarm sounds on a fireground, signaling that you're in the "uh-oh" zone—in other words, time is critical. Your entire world narrows to that sound, and your focus locks in on every breath you're taking. You become acutely aware of how quickly your air is running out, and the urgency pumps your adrenaline higher. This is where regular training on air helps you to stay calm, control your breathing, and navigate the additional challenges posed by anxiety.

If you train like it's the real deal, then you know how to work around breathing too fast by slowing yourself down. This skill is learned through time, effort, and experience—not through luck. Your heart's pounding so hard it feels like it's going to break through your chest, and your thoughts are spiraling out of control: *What if I can't make it out? Did I train hard enough? What did I do wrong? Are my crew OK? What if no one comes for me?*

All you hear will be the alarm, your heartbeat, and your own gasping breaths. The anxiety is suffocating. You're trying to calm yourself, but it's terrifying, like trying to stop a runaway train. You're standing there, in the middle of all this chaos, and suddenly you're a scared kid again, wishing someone would come and help. You feel like you're drowning, with no one to rescue you.

You're stuck, lungs burning, mind racing, and every second that goes by feels another step closer to disaster. It's a full-blown panic attack, but you can't run or hide. This job has a way of stripping you down to nothing by exposing every weakness and flaw. That moment was a wake-up call: I couldn't be the weak link anymore. I couldn't be the one who ran out of air first, who wasn't ready when it counted. That sheer panic became my motivation to do better, to be better. I was determined never to let this happen again. In such moments, you see who you really are and how badly you want this job.

Reflecting on the call/volunteer academy, I had a poor attitude at the time, never taking air management seriously. I was lazy and unmotivated, thinking it was enough to show up. "Good enough" doesn't cut it, though. It's not about being able to pull a hose or swing a sledgehammer; it's about controlling your breathing when your body is screaming at you to stop, pushing when you're tired wanting to quit, and staying calm when everything around you is a living nightmare.

Key Takeaways

Air is life and comes at a cost. Every breath on air could be your last in a deadly environment. Take nothing for granted.

Your SCBA isn't just gear; it's your lifeline. Respect it. Train with it like your life depends on it, because one day it will. Fitness isn't optional; it's key to survival. Your bottle will expose your preparation—or lack thereof. Stop making excuses.

Panic is your biggest enemy. When your low-air alarm sounds, mindset is everything. If you lose your head, you lose your way out.

You're only as good as your last breath. Complacency kills. You have to show up to every shift ready for the worst because one day, the worst will come.

You don't rise to the occasion; you fall to your level of training. The fireground isn't the place to wish you had worked harder.

"Good enough" will get you or your crew hurt. The fire doesn't take excuses. The job demands your best.

Air management is a skill, not luck. You don't master it by showing up. You master it by putting in the work, including when nobody's watching.

Train like it's real because one day it will be.

January 2024

In my career department, I knew things had to be different, so my mindset shifted. We did similar air consumption drills as before, but this time I was no longer proving I was the best; it was about survival, making sure I didn't run out of air when it mattered most. I wasn't competing with anyone but myself. I made that bottle last nearly 30 minutes, learning how to control my breathing and mind in the middle of commotion.

Final Thoughts

You don't want to be the reason why one of your crew comes close to dying because you didn't put in the work and couldn't control your breathing. If you can't help yourself, how can you help any of your firefighters? That's why I take this job so seriously and push myself in every workout, every drill, and everything I do—because one day it won't be a drill. When that day comes, you need to know that you're prepared. Train like your life depends on it, because it does. When the real deal happens, you'll thank yourself for taking air management seriously.

Reflection Prompts

- ✠ Describe a moment during a drill or scenario when you realized how critical air management was.

- ✠ Where do you struggle most with air management: panic, pacing, or physical conditioning? Be honest.

- ✠ What's one mistake you made with air management that taught you something valuable? Would you do it differently, or are you glad it happened?

10

FIRST REAL FIRE: WHAT DO I DO?

February 7, 2024, 0736 hours

At the station, ready to relieve the previous shift, adrenaline flooded my system as the tones dropped: "Vehicle fire extending to the structure." Acknowledging the report, the captain assigned both engines, the ladder, and the medic, with mutual aid requested.

The station exploded into motion—boots flying, turnout gear donned in seconds, and everyone scrambling to get on a truck. The captain shouted to me, "You're driving the engine. Hop in!" I was behind the wheel before my brain had fully caught up with the situation. My thoughts were a chaotic mix of nerves and excitement, amid the lights and sirens.

Arriving second due on scene, the ladder crew stretched the 200-foot cross lay and tagged the hydrant. My officer ordered me to take the engine to the other side of the structure: "Go around, go down the next street, and hit the other end." With no hesitation and no questions, I set the truck in place, tagged up on the accountability cone and air pack, and teamed up with the ladder crew. Then the chief's order came through the chaos: "Get on the line and head inside with us."

Pure excitement radiates through my body, as I think to myself, *No way, dude. I'm going in!* .

Masked up, ready to go on air, I'm feeding the hose as my crew makes the initial entry. There were only four of us, two on each piece. The initial cross-lay attack line was a 100-foot walk to the rear of the house, leaving 100 feet left to work with; however, it was a two-story residential, so doable. The ladder driver, assigned as pump operator, charged the line. I'm the third crew member, feeding the line into the house.

Now the hose becomes wedged between the railing and the siding while the captain is advancing into the two-story dwelling. I radioed the complication to the captain, and we did what we could with what we had, despite our minimal staffing. A fire doubles in size every 30 seconds, so we didn't have time to grab another line. I clipped my regulator into my SCBA facepiece and advanced into the building, following the hoseline through the smoke.

Remember Your Training Foundations

Looking for any sign of trouble, I made my way toward a very narrow stairwell. Meanwhile, the voice in my head was loud: *Keep your head on a swivel; do that size-up.* Seeing fire extension on the wall just above my head, I called out to my captain, informing him of the conditions.

Then, like a genius probie, I froze—standing there waiting to be whacked in the head with common sense. I was mesmerized, caught by tunnel vision, staring at the flames like they were something out of a dream. I knew better, but my brain had short-circuited. Completely unaware that the third engine crew had arrived, I hear a familiar voice holler, "Get up there now!" It's Sterls, my friend who got me the career position, who I've worked with since I started in May 2021, and as he joins me, we boogie up the stairs.

At the top of the steps, intense heat slams into us. Sterls and I drop to our knees, flames rolling over our heads. The stairwell is now an active chimney. We are backed into a wall at the top of the stairs with nowhere to go, only one line stretched already advanced way ahead of us. I have zero clue of the layout of the house, my hair is frayed, and I'm staring at the last bit of flame I can make out through the dark heavy smoke.

It had turned into a complete blackout, with zero visibility. We went from not great to really bad in less than 30 seconds. Fear didn't grip me. Instead, a cold, hard truth settled in: There were no instructors to help me, no one to inform me I'd made a mistake; I was on my own and had to trust my training and my crew on the line with me. Then suddenly came the welcome sounds of water spraying and glass shattering. The nozzle operator had made his way to us and knocked the flame down into complete extinguishment.

Originally, this fire began with a remote starter for a vehicle that was parked in the driveway. Once the vehicle was ignited, the fire extended to the house rapidly. From there, it got into the siding through conduction and made its way into the small crawl space, which led to the fire extension in the stairwell.

The fire extension above my head that I had called out to the captain turned out to be above a small hatchway door. By the time Sterls and I got to the top, the hatchway, pressurized from the heat and off-gassing, had already blown open, creating the inferno that had made the stairwell seem like a chimney.

After extinguishment, Sterls and I overhauled the crawl space, searching for any hidden fire extension. This entailed about 15 minutes of moderate overhaul, making sure the fire is completely out, eliminating hot spots, and looking for any other fire hazards. Pike poles smashed through the drywall, and sweat poured down my face. More mutual aid crews arrived shortly after our chief relieved us from overhaul.

My heart pounded and my mind raced. This was my first structure fire—a significant moment for me, even if it was just another day on the job for the seasoned firefighters.

Lessons Learned

The academy doesn't knock you down for the sake of knocking you down; it's about making sure you can do the job efficiently—and survive. This job can kill you in seconds, and once you walk out of those academy doors, there's no one to hold your hand. It's on you to remember what you've learned to keep yourself and your brothers and sisters safe.

We love this job. When it's time to go to work, we get it done, no questions asked. I don't wish structure fires on anyone, because in the end, it's someone's home and life going up in flames. But when the call comes, we don't hesitate.

My department isn't like those in big cities. From day one, we're trained to be independent. With only six on a shift, we're lucky to have everyone there when a big call comes in. We run two ambulances on top of the fire apparatus, and if both are out, the station is understaffed. Emergencies don't wait for a full crew complement.

When crap hits the fan, it's going to fly. And you need to be ready—no excuses. Train during downtime because your life depends on it, and so do the lives of your crew. That's the reality of this job. You're not just a firefighter; you're the difference between life and death every time those tones drop.

Just 12 days out of the academy, I stared down my first structure fire. Fire's don't wait for you to be ready; they come when you least expect it.

The instructors taught us how to not get killed. Once you're on your own, there are no slaps on the wrist—one wrong move and you or someone else is dead. The academy doesn't hand out deficiencies for kicks; it's about survival. Take what you learned and never forget it.

Key Takeaways

Your first fire will expose everything you didn't realize you weren't ready for. It's where training turns into instinct—or doesn't. No matter how chaotic it gets, you have to slow your mind down, fall back on fundamentals, and communicate clearly. Tunnel vision is real, and so is hesitation. But both can be overcome with preparation and awareness.

Fire grows fast, staffing is limited, and real-life conditions don't wait for textbook scenarios to play out. In a minimally staffed department, there may be no second line, no perfect layout, and no one walking you through it. You are expected to operate, think, and contribute like everyone else on scene because the fire doesn't care who the probie is. The lesson is simple but critical: Respect the job, respect the pace, and respect your training. When you mess up, own it. When you're unsure, speak up. And when it's your turn to act, you act.

Being ready isn't about having all the answers. It's about being present, being useful, and giving 100% even when the situation isn't ideal. That's what keeps people alive. And that's what earns you trust on the line.

Final Thoughts

Your first fire is a moment you'll never forget—not because you'll do everything perfectly, but because it will teach you exactly where the gap is between the classroom and the fireground. It will humble you. It will show you what fear, responsibility, and pride feel like all at once. There will be no instructor to grade you and no reset button—just you, your crew, and the fire. And in that moment, you will realize what this job truly demands—that you trust your training, own your decisions, and keep moving forward, even when your mind is racing.

You won't be perfect. None of us are. But if you stay humble, ready, and focused on the team, you'll come out stronger on the other side. The fire doesn't care who you are, but your crew do. And they need you to be all in.

Reflection Prompts

- What was your first fire like?

- What are some lessons you learned from your first fire?

11

UNDERSTANDING FRICTION LOSS

May 31, 2021

I had officially been thrown into the abyss of the fire service. I was shown a pump panel with no clue what I was looking at. Confused and terrified, I realized this job isn't just about showing up and putting out fires; you actually have to figure things out logically. It may not come naturally, like learning a foreign language, and you may be left standing there feeling stupid because you don't understand it.

I was ready to give up the first time I looked at a pump panel while someone tried explaining what friction loss was and why it was important. My head felt like it was going to explode. But working in a small-town department—where we run with minimal staff, six or sometimes only five of us on shift—I had to know everything. I couldn't rely on someone else to handle the weak points for me. This is my job now, and there's way more to it than I ever realized. There's no faking it—either you know it or you don't.

During the full-time recruit academy, it hit me that I needed to understand friction loss better. Then one day in the drill yard, it just clicked. Realizing I could understand something so confusing as friction loss, I started to gain the confidence I carry now. And that's exactly my point: If you're feeling overwhelmed or confused, know that it's normal; understanding comes with time. Don't beat yourself up for not getting it right away. You'll get there.

The Basics of Friction Loss

In the simplest terms, *friction loss* is reduction in water pressure, in pounds per square inches (psi), as the water moves through a hose. The longer the hose or the more water you're pushing through, the more pressure you lose. Think of it like this: If you're trying to suck water through a long straw, it's going to be harder to get the water to move efficiently the farther away it is. The same thing happens with water flowing through a fire hose. Friction loss calculations aren't as hard as the academy makes them seem, though! This is fireground hydraulics, not advanced algebra.

Occasionally, I refer back the friction loss cheat sheet I followed when learning to pump at the academy (fig. 11–1). It really is simple: First, you figure out how many feet of hose you have and what appliances you're using; next, you add it all together. Use the skills or lose the skills.

This cheat sheet might look like a lot, but with practice and experience, it'll eventually click and make sense. The terminology from your lectures isn't as important as hands-on practice. If you're in a small town, rarely will you charge more than two lines at once.

RULE OF THUMB - FIREGROUND FRICTION LOSS

1½"

FLOW	FL/100'
50 GPM	10 PSI
80 GPM	20 PSI
100 GPM	30 PSI
125 GPM	50 PSI

2½"

Drop the last digit of the flow, subtract ten from that number.

Example: 25ϕ GPM
-10 GPM
15 PSI/100'

(Above 399 gpm, just drop the "0")

5"

FLOW	FL/100'
500 GPM	2 PSI
1000 GPM	5 PSI
1250 GPM	10 PSI
1500 GPM	15 PSI
2000 GPM	20 PSI

1¾"

FLOW	FL/100'
100 GPM	15 PSI
125 GPM	25 PSI
150 GPM	30 PSI
180 GPM	40 PSI
200 GPM	60 PSI

3"

Determine the flow in hundreds of gallons per minute and square the first digit.

Example: 300 GPM

3 x 3 = 9 PSI/100'

Efficient Carrying Capacity

1½"	100 GPM
1¾"	150 GPM
2"	200 GPM
2½"	300 GPM
3"	500 GPM
4"	1000 GPM
5"	2000 GPM

2"

FLOW	FL/100'
125 GPM	10 PSI
150 GPM	15 PSI
200 GPM	25 PSI
250 GPM	40 PSI
300 GPM	55 PSI

4"

FLOW	FL/100'
500 GPM	5 PSI
700 GPM	10 PSI
750 GPM	12 PSI
1000 GPM	20 PSI
1250 GPM	30 PSI

STATIC PRESSURE: Pressure on the compound gauge with no water flowing. The total amount of water available.

RESIDUAL PRESSURE: Pressure on the compound gauge with water flowing. The remaining water available

Parallel Lines

2 - Parallel 2½" Lines

Divide total flow in half, calculate flow on one line.

3 - Parallel 2½" Lines

Divide total flow in thirds, calculate flow on one line.

Static/Residual Rule

Pressure drop of 0 to 10%:
3 times the present flow available
Pressure drop of 11% to 15%:
2 time the present flow available
Pressure drop of 16% to 25%:
1 times the present flow available
Pressure drop of more than 26%:
Minimal flow available

ALWAYS MAINTAIN 20 LBS RESIDUAL

Flow From Solid Stream Tips

Hand Lines

7/8"	150 GPM
15/16"	180 GPM
1"	200 GPM
1 1/8"	250 GPM
1 1/4"	300 GPM

Pump Capacity - net pump pressure

100% of pump capacity @ 150 PSI

70% of pump capacity at 200 PSI

50 % of pump capacity at 250 PSI

Nozzle Pressures Hand Lines

Solid Stream	50 PSI
Traditional Combination	100 PSI
Low Pressure Comb.	50 PSI
Low Pressure Comb.	75 PSI

Line Pressure equals nozzle pressure (NP) plus friction loss (FL), and adjustments for back or forward pressure and/or appliance loss.

Flow From Heavy Stream Appliance Tips

1 1/4"	400 GPM
1 3/8"	500 GPM
1 1/2"	600 GPM
1 5/8"	700 GPM
1 3/4"	800 GPM
1 7/8"	900 GPM
2"	1000 GPM

1 - 2½" and 1 - 3" Line

Divide total flow in half, drop the last digit, and subtract 15.

Example: 500/2 = 25ϕ
-15
10 PSI

Master Streams

Solid Stream Nozzle	80 PSI
Combination Nozzle	100 PSI
Master Stream AFL	25 PSI

Non-Master Stream AFL (appliance friction loss) ie: wye, siamese, etc.

0 GPM to 349 GPM	0 PSI
350 GPM and above	10 PSI

Back (or forward) Pressure

Add 5 PSI per 10' above grade.
Subtract 5 PSI per 10' below grade.
(Each building story is appox. 10')

"Rule of Thumb"
Fireground Hydraulics

[MFA - Rev. #6 – 8/16/11]

FIG. 11–1. Friction loss cheat sheet (Courtesy of the Massachusetts Firefighting Academy)

The cross lays we carry on our trucks provide a practical example from my department. A *cross lay* is basically the hoseline you'll grab first to attack a fire. It's preconnected to the truck, coiled or stacked. So it's easy to pull off fast, and it usually sits right above or beside the pump panel. You pull it across the truck (hence the name), you stretch it out, and you're ready to go. Simple as that—it's your go-to attack line for most fires. The reality is this: You're pulling a line, charging it, and getting to work. The nozzle operator will let you know if more pressure is needed.

Need-to-Knows

Here, I'll explain simple friction loss with basic cross lays, so hopefully it make sense to you. I teach individuals based on what works for my brain. I figure there's enough like-minded individuals who learn in the same way. So bear with me as I do my best to explain this to you without a million equations. Be the sponge that absorbs this information. Remember, this is just the bare minimum to help you go from not knowing what friction loss is at all to mastering it intuitively. I'll try not to make you yawn.

We start with *hose sizes*. These are the two most common hose sizes for any piece:

- 1¾-inch hose: 30 psi friction loss per 100 feet
- 2½-inch hose: 15 psi friction loss per 100 feet

You'll use 1¾-inch hoses more often than anything else as your attack lines. The 2½-inch hose delivers big water for big fire. Make the job easier by pulling the right line; don't yank the 2½-inch hose unless you need it. If your probie pulls the 2½-inch hose needlessly, that's on you for not training them correctly; now you've got your work cut out to retrain them properly.

Then we figure in the *nozzle*. The most common nozzles are the *combination tip* and the *smooth bore*, corresponding to the following nozzle pressures:

- Combination tip: 100 psi
- Smooth bore: 50 psi

Line pressure is the sum of hose friction loss plus nozzle pressure. Let's say you have a 1¾-inch cross lay with 200 feet of hose; that's 60 psi (i.e., 30 psi for each 100 feet). Now add in your nozzle pressure; for a combination nozzle, that's 100 psi. So, 100 psi from the nozzle plus 60 psi from the hose friction loss gives you 160 psi line pressure.

For the 2½-inch hose, it's the same process. Since there's 15 psi per 100 feet, it will be 30 psi for 200 feet of hose. Add 50 psi for the smooth-bore nozzle, and you've got 80 psi total.

Thus, the key to pumping a line is knowing your hose length and nozzle type, then adding together the hose friction loss plus the pressure of the nozzle used, usually combination tip or smooth bore. The math doesn't change based on the truck—only the nozzle and the efficient pumping capacity of the line.

The Basics

Tank Fill

We all start out as beginners. First, put the truck in "pump." Then, open your tank to "pump" and partially open the *tank fill*. Finally, open the line that's been pulled off the truck by the nozzle operator—but only once they call for water. Pay attention not to charge the line until they're ready.

Being proactive as a pump operator means clearing the hosebed of the lines that were pulled and making sure that the lines are free of kinks and won't get hung up in a bad spot. Taking the extra 30 seconds to look at the line your crew pulled off the truck can make the difference between a smooth operation and one that has to be paused—and could even be career ending. So, give a damn and pay attention.

After the nozzle operator indicates that they're ready for you to charge the line, open the valve slowly. A walking pace is ideal because it will be much less abrupt for the nozzle operator. Then, throttle up to your line pressure. You do not need to use your tank fill to charge a hose line.

For reference, the two most common reasons to open your tank fill are

1. To keep water circulating and the pump cool so you don't burn out the pump when the truck is sitting idle, and

2. To refill your truck's tank from an incoming water source.

You don't need the tank to fill open the whole time while pumping, but it's a good habit to use it if you're going to be sitting in the pump idle for a while. Once your line is established, manage the tank fill as needed and pay attention to your water level so you can keep it from dropping.

Elevation

For every 10 feet, add or subtract 5 psi from your line pressure. If your crew goes up 2 floors (each floor is about 10 feet), add 10 psi (i.e., 5 psi per floor). If they go down, subtract 5 psi per floor. Easy enough.

Hydrants

Residual pressure and static pressure are distinct quantities. *Residual pressure* is the minimum pressure remaining in a system when water is flowing, typically measured during peak demand or when a hydrant is in use. *Static pressure* is the maximum amount of pressure you can expect in a system.

You really need to see this in person, with a pump panel in front of you. Every truck is different, and some towns—my call department, for one—don't have hydrants. Drafting with drop tanks and water shuttles might be your only option (your incident commander knows this already and should be ahead of the game).

Pay attention to your master intake gauge; residual should always have 20 psi. That number determines how much you can do as a pump operator efficiently without causing any possible catastrophes.

Friction Loss Tips

Know your trucks. Know where everything is, including cross lays. And know your line pressures.

Make a cheat sheet, laminate it so it won't get ruined, and keep it where you can quickly access it. Ask the members of your department what line pressures they run normally if you don't know. Hopefully, all your cross lays are the same length, which makes the math easier. When I started out, I did the calculations and used a label maker to print every friction loss and stick it right on each panel (some are still there almost 2 years later).

When pumping, pull out your cheat sheet, and pull the correct cross-lay valve. (Double-check the cross lay before charging it to make sure you charge the right line, not the one still in the bed. I say this because it's been done—and no, it wasn't me.) Then, charge the line. Don't be the probie who pulls the valve on the cross lay that's still stacked in its home in the bed of the apparatus. Throttle up to the right revolutions per minute for the line pressure in psi—and just like that, now you've got water to the fire.

As a final tip, be a proactive pump operator by checking your hosebed before you charge the desired line to make sure that all of the cross lay has been pulled out.

Final Thoughts

When you're new, friction loss and pump panels can make your brain feel like it's about to short-circuit. I get it—I've been there. I stared at that pump panel like it was from an alien spaceship, and I wanted to throw in the towel more than once!

Key Takeaways

Friction loss is reduction in water pressure as it moves through the hose. The longer the hose or the more water you push, the more pressure you lose.

The 2 most common hose sizes are 1¾-inch hose, with 30 psi friction loss per 100 feet, and 2½-inch hose, with 15 psi friction loss per 100 feet. For nozzle pressures, combination tip will add 100 psi, and smooth bore will add 50 psi.

The line pressure formula is as follows:

- Hose friction loss + nozzle pressure = line pressure.
- For example, a 1¾-inch cross lay at 200 feet with combination nozzle will have a line pressure of 160 psi.
- Elevation adjustment requires adding 5 psi per floor up and subtracting 5 psi per floor down (every 10 feet).

Pump operator basics are as follows:

- Put the truck in "pump."
- Open the tank to "pump," partially open the tank fill, and charge the line slowly and smoothly.
- Watch water level and residual pressure (keep 20 psi or more).
- Be proactive about addressing kinks, clearing hosebeds, and pulling cross lays.
- Know your truck.

Remember: Cheat sheets are your friend. Laminate one, label your panel, and double-check your lines before charging! Finally, don't overthink it. Fireground hydraulics isn't rocket science—it just takes practice. And that's what turns a rookie into someone worth relying on when it really counts.

The truth is, you will get it. It just takes time, repetitions, and actually putting your hands on the equipment. This stuff doesn't click overnight, and that's OK. Don't beat yourself up, and try not to feel stupid. Every single firefighter, no matter how salty they are now, was confused about this at first.

The key is to pay attention, take the time to learn it right, and care enough to slow down and think. You'll go from standing there with a blank stare to confidently working that pump. Stick with it and you'll have it down in no time!

Reflection Prompts

- Do these tips make calculating friction loss easier to understand? If not, where are you getting hung up?

- How does your company use the pump operator/driver position?

12

The Chirp

Firefighters are jacks of all trades. Not every call is a fire, a medical emergency, or even a cat rescue. Sometimes it is a public assist because of flooding, keys or ducklings stuck in a sewer, or a chirping alarm that no one can find. No day on this job will be the same as another, and it pays off to learn things that initially seem irrelevant—like how to start a snowblower or disengaging the chain brake on a chain saw before taking the moving parts cover off. These skills that you never thought you would need will come in handy.

How great are you going to feel if your next run is for a public assist to start a snowblower, and you don't even know to turn the ignition switch? Or, if you have a new member reporting to you who decides to explore the chain saw, how smart will you feel when they ask for help putting it back together, but you can't, because the chain brake is still engaged and you have no idea why? Then, a structure fire comes in and you don't have a working saw to vent the roof anymore? All because you didn't take the time to go through simple tasks that take 5 minutes to learn. The small things matter.

February 5, 2024

We had been dispatched to a commercial property for the report of an alarm chirping, because the occupants couldn't figure out where it was coming from. This mysterious chirp had been haunting the frazzled property owners for days, driving them to insanity. Now, it was our problem to find the source.

Gas meter in hand, my partner, who had over a decade on the job, ordered me to meter the basement and work my way up each floor. We kept our ears peeled for the elusive chirp. The occupants swore it was there, but of course, the moment we arrived, the chirp was silent. But just as we began to doubt our sanity, the chirp came back—faint, but still there. The hunt was on!

The chirp was an aggravating tone, somewhere between a dying battery and a carbon monoxide detector. We moved through the rooms of the home, checking above drop ceilings, listening at every turn. Eventually checking the basement and attic off the list of possible sources, we narrowed the source of the annoying sound down to the first floor. An hour in, we were hot on the trail but still hadn't located the elusive sound.

The chirp seemed to be mocking us from within the walls. My partner asked the owners if they had done any renovations, speculating that a faulty detector could have gotten buried in the walls. There had been no recent renovations, though, at least not that the current owners recalled.

Now I was determined. Pacing back and forth, I pinpointed the sound to a spot on a wall between a hallway and one particular room. The chirp grew louder, still taunting me invisibly. Low to the ground, near a desk in the hallway, it was deafening! “It’s gotta be here,” I swore, sticking my head under the desk. The chirp was blasting in my ears.

A recycling bin under the desk didn’t seem suspicious until it did. My partner grabbed the bin, and the source of the chirp finally was revealed: In a plastic bag within the recycling bin were three hardwired detectors and alarms, with batteries still in each. Turns out that the occupants had moved into the building relatively recently. They’d been updating the place, replacing alarms, and had tossed the old detectors into the recycling bin, promptly forgetting about them. They appreciated our help, and ultimately we left them in peace, if also a little embarrassed.

Such calls are one in a million. We’re firefighters, but we’re also the Swiss Army knife of public service. We do it all. If someone calls for help, you can bet we’re showing up. As ridiculous as some calls might be, we’re the solution to 99% of the public’s problems. And if we can make the problem go away or at least rule out an emergency, then we’ve done our job, so appreciate the calls you find irrelevant.

Key Takeaways

The job isn't just about fire and EMS calls. We're called for everything from fires to chirping alarms to ducklings stuck in drains.

Master the small stuff. The 5 minutes you spend learning how to start a snowblower, reset a chain saw, or troubleshoot an alarm might save the day and bolster your credibility when it counts.

Every call matters. The ones that feel silly or irrelevant are often the ones where we build trust with the community and sharpen skills that could matter later.

Patience and determination pay off. Whether you're chasing a chirp or figuring out a tool, don't give up.

Stay humble. Everyone has those moments when they feel like an idiot. Our job is to help, not judge.

Final Thoughts

It's easy to think we're here for the big ones—the fires, the rescues, and the dramatic calls that make headlines. But the truth is, we're here for all of it. The small calls, the weird calls, and the calls that leave you shaking your head are just as much a part of the job as the big ones. These moments teach us to stay humble, pay attention to the details, and always be ready for whatever could come next. So the next time you find yourself chasing a chirp or figuring out how to start someone's snowblower, remember that this is the job. This is what service looks like. And if you can laugh about it later and walk away knowing you helped? That's a win.

Reflection Prompts

- ✠ What's the most unexpected nonemergency call you've been on?

- ✠ Name a skill that you didn't think you needed but developed during your time in the fire service.

13

Come On, O'Donnell!

February 12, 2024, 1800 hours

Fall-off-the-bone ribs slathered in Sweet Baby Ray's barbecue sauce, so good I'm grinning with the mess embedded into my smile. A side of cheesy garlic mashed potatoes creamier than the ice cream waiting for me afterward. Crispy Caesar salad and corn on the cob so hot it'll have me screaming "Someone call the fire department!"

Plates are set. Croutons crunch as the shift digs in. My cooking is edible. I know it's delicious when the table's dead silent, no one daring to speak lest they break the spell. Then, nothing screams probie like taking that first sip of my drink, placing it on the table, and knocking it over. "Come on, O'Donnell!" Right into the laps of three grown men, the captain included.

Now, thanks to my clumsiness, three seasoned firefighters look like they wet themselves. *Good going, probie!* Sippy cup once again issued to me. That's one way to make a great impression.

First on Scene

February 24, 2024, 1451 hours

We've been dispatched to an reported electrical fire with smoke in the basement. The homeowner says it smells like burning plastic. Type 5 lightweight construction, two-story dwelling on a downhill grade—the kind of place that practically begs to go up in flames if you so much as sneeze at it. The ladder, engine, and medic are on the way, and lucky me, I'm riding with the ladder crew. We're the first due company.

Helpful Insights

Our town is split into two districts: north and south. There are two stations, staffed by six firefighters total. Headquarters handles the north, and Station 2 handles the south.

Some areas have hydrants, while others are basically a firefighter's nightmare. In those unlucky spots, you're *drafting*. This is when you create a vacuum seal through gravity and physics. There are two types of hose: soft sleeve and hard suction. Hard suction is for drafting; soft sleeve is for fire attack, sometimes supply.

You connect the hard suction hose, which is a rigid hose with gaskets inside. Importantly, make sure the gaskets are present, so that you don't have any leaks. Connect the hose to your strainer and drop that into your water source, then connect the other part to where the intake is on the pump; this is where you normally connect your hose to receive water from the hydrant.

On our trucks, we have to unscrew the intake off the pump and swivel the hard suction hose onto the pump panel manually. We make sure everything is tight and no valves are open; otherwise, you won't get a vacuum seal or any pressure. Then pull your prime, if you can get one, and pull water from a drop tank, pond, or cistern. Now you've got yourself a makeshift hydrant—gravity and physics defiant.

Graceful Descent

We're greeted by two fluffy huskies. They're barking loudly, as if we were going to steal their toys, while we make our way inside to meet the homeowners. We're led into the basement, down stairs that are bright orange with fire covered in enough dog fur to outfit a small army.

At the bottom of the stairwell is a concrete slab. As we make our way through the basement, it smells like a plastic factory. Thick, white smoke is everywhere. We've got a thermal imaging camera (TIC), irons tools, and a water can with us. Now, we're hunting down the source of the smoke.

The culprit is a century-old crusty-dusty shop light held up by chicken wire. The TIC reads 180°F on this prehistoric artifact. (When asked about it later, the homeowner replies, "I didn't even know that thing was still there, let alone hot.") As the basement is turning into an inferno, my captain reaches the panel, kills power to the affected area, and assigns me to grab the cutters from the tool bag. In hindsight, I should've had that tool in my coat, but I was too new to know better.

I head out to the rig, grab the cutters, then march back into the house. After one step, I lose my footing and gravity decides to take over. Fifteen thuds and a final grunt later, I land on the concrete slab with an abrupt stop. Laid out flat, wondering if someone caught the license plate of the truck that hit me, I'm hoping no one saw my fall. As it happens, both my captain and the homeowner witnessed it and are trying to figure out if they should laugh or call 9-1-1, except that I am 9-1-1!

Standing back up, I hold the tool up high and proud. "Here's your tool, Captain!" Shaking his head, he cues me to cut the fixture down as he informs the homeowner to call an electrician (don't mess with the wiring, and don't turn the power back on until someone with more qualifications than a YouTube tutorial checks it out). I grab the shop light, haul it upstairs, and toss it into the front yard like it's a ticking time bomb.

I'm grateful for extra cushioning in my tush that acted as a crash pad! I gave everyone a free comedy show and earned myself a starring role in the next Human Resources meeting. (And the huskies were unharmed, by the way.)

February 26, 2024, 2249 hours

It's a bone-chilling winter night, with everyone tucked in bed except the captain, lost in his latest novel in the dayroom recliner. Floors mopped, dishes washed, coffee maker preset for the morning; the only thing left is my sacred nightly constitutional, with 20 minutes of blissful peace and quiet.

Or so I thought. Once I decide to wrap things up, I stand—and whack! Abruptly, I fall back on the toilet, not by choice but by the sheer force of my stupidity. There's a cabinet right above the stall, in the worst spot for someone with the audacity to be nearly six feet tall. The cabinet corner is sharper than my memory. So there I am, sitting with my head throbbing, contemplating my life choices.

After a more careful second attempt at standing, I shuffle out of the bathroom after having some sense knocked into me. Only then do I feel a suspicious warmth trickling down my face. I go back to the bathroom mirror and sure enough, bright red blood is pouring down my face as if I'm auditioning for a horror movie. Suddenly, I'm on the fast track to a full-blown freak-out. Now it's time for the dreaded walk of shame into the dayroom.

The captain's just trying to enjoy his book, not expecting his night to involve any probie clumsiness. Tucking my tail between my legs, with a hand on my head, I it spit: "I just split my head open on that stupid cabinet. Got anything to stop the bleeding?" Currently I'm convinced I'm going to need stitches—and maybe a new brain. Standing there like a lost puppy, trying not to look pathetic, all I can think about is the mountain of paperwork and the reputation I'm about to cement, hurt within the first year, in the dumbest situation possible.

Our company has an advanced life support (ALS) setup with every imaginable medical supply—including bandages and other materials to stop bleeding. Guess who didn't remember that, though, in the panic of her bloody incident? The captain first asks if I'm OK and then hands me some paper towels, tells me to keep pressure on it, and heads off to grab some real supplies. He returns with a 4 × 4-inch gauze pad, and after what feels like forever, the bleeding finally stops.

No stitches are required, leaving me with just a gnarly scab and a story no one would believe if I didn't take a picture (fig. 13–1). The captain gave me the option to report the injury, but I refused to fill out any paperwork. I know my head better than anyone else. After all, I've just tested its durability. The human head's got a lot of blood vessels, so it's normal to bleed like a fountain when you crack it open. As I stood there, replaying the whole disaster in my

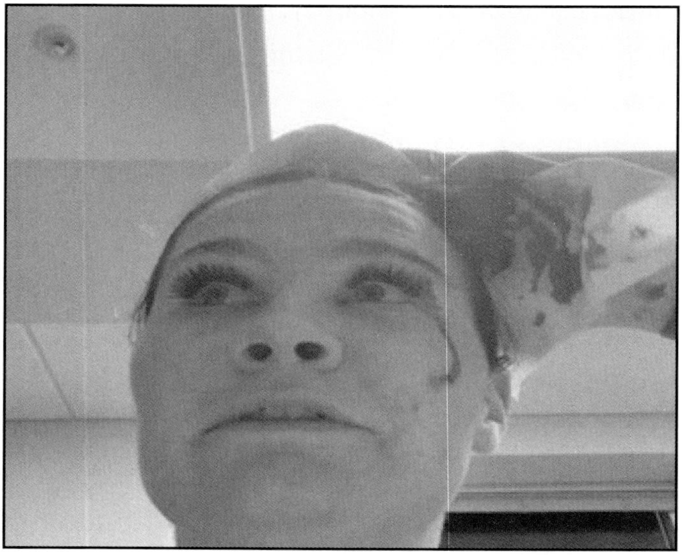

FIG. 13–1. Photographic evidence of what happens when a head collides with a metal cabinet

mind, I couldn't believe I'd asked, "Do you have anything to stop the bleeding?" We're in a firehouse that's practically a minihospital—as some call it, a "doc in a box."

I mop up my trail of shame from the bathroom to the dayroom and can't help but laugh at myself. At this rate, my captain deserves a promotion just for the headaches I cause—literally and figuratively.

March 23, 2024, 1830 hours

Dinner time at the firehouse. Typically our shift splits the cooking duties evenly, but as the probie, I've taken it upon myself to whip up homemade meals for the crew. That's what's expected in your first year, right? But today, things were different. Call volume was higher than average, so we decided to be sensible and order Chinese food instead. One of the crew called ahead, and I'm delegated to pick up the order in our squad vehicle.

Arriving at the restaurant, I nearly avoid getting smoked in the parking lot by a driver who's more interested in texting than watching the road. I don't think I've ever heard brakes squeal that loud. Gotta love distracted drivers! Anyway, he stops in the nick of time, and I head inside to pick up our dinner. At the counter, the clerk gave me a discount, which was thoughtful. I thanked her, saying it wasn't necessary, but she insisted. So far, so good, right? I head back to the station, pull toward the bay, and make sure the garage door is fully open.

Here's where I pull the probie move. I back up, stop, and do my usual check to see how much room I've got. Reflexively, I push the button to close the garage door button while I'm still checking. Before I can trigger the sensor and stop the door from closing, the door taps the hood of the squad, stops, and then begins to go back up. I'm frantically pushing the stop button, but it keeps going.

For those of you who don't know, once a garage door hits something and goes back up, you've got only a few seconds before the safety string pulls and snaps the safety arm. It's essentially the backup emergency system. The door gracefully tapped the hood, with no damage, thankfully, but guess what? I was too slow, and unfortunately, the safety arm snapped.

Sheepishly, I walk into the dayroom, place the food on the table, and confess: "Cap, I screwed up. I sent the garage door into the squad hood, broke the bay door, and snapped the safety lever."

After investigating, the captain shakes his head and reassures me that it's all right. I replied, "Well, I'd rather tell you I screwed up and own it instead of pretending it never happened and hoping no one finds out."

Then, he tells me, "You're not the first one to do that. Someone did the same thing 3 days ago." Turns out, the sensor didn't catch the door because the bumper sits 3 inches above it. No wonder it thought that nothing was there. One of the members chimes in, "Don't feel bad. I shattered all the bay windows one time with the engine."

Owning It

I had to address the incident and sent an email through the chain of command. After I did so, just like that, it was over. The safety lever was fixed the next morning, and the door was still operable manually meanwhile. All it needed was to be bent back into place and restrung. The bay door was fixed like nothing had ever happened. Ultimately, you can't avoid every mistake. Own up to it when you make a mistake, and it'll work out.

Take extreme ownership of your actions and face the consequences, which are never as bad as your mind makes them out to be. Stop stressing, take a deep breath, and learn from your mistake. Don't let it become a habit.

Everyone makes mistakes. I guarantee I'll keep making them, which just means I'm learning—*all of the time.*

March 31, 2024

It's an ordinary day at the firehouse. I'm doing my probie duties, trying to stay busy while the radio is silent. You've got to find something to do during downtime; I don't care what, but you have to keep busy. I'm a believer in the use-it-or-lose-it philosophy, and my default is to practice knots whenever I can. On this shift, I decided to engage in one-on-one training—me and my knots versus my trusty 210-pound dummy.

Today's exercise was all about the California love knot, also known as the handcuff knot, which sounds way more fun than it actually is. This knot keeps flailing arms under control, which is great for rescuing an unconscious firefighter—or whatever else you might need it for. No judgment here if you're into that.

Tying up the dummy, I lift, drag, and carry it around the apparatus floor, imagining all the heroic scenarios where I might need these skills. Getting this dummy into a firefighter carry was an absolute beast of a time. Finally, I got it over my shoulder and started squatting. Squats with weights? Easy! Squats with a 210-pound deadweight dummy? Way different.

On the last repetition in my series, I hear an ominous ripping sound. The seam of my pants had popped and split up the middle, right in the apparatus bay, of all places. Dropping the dummy faster than a hot potato, I stood there in shock, hearing SpongeBob SquarePants singing "I ripped my pants" on a loop in my head. And for posterity (pun intended), I was recording the entire exercise to check my technique later!

Accepting Defeat

Strutting—or perhaps doing a walk of shame—over to my captain, I fess up: "I don't know how else to say this, but I split my second-to-last pair of station pants doing squats with the mannequin." I'm trying to keep a straight face as I giggle because, let's be honest, my tush is practically a national treasure at this point.

Can you guess why this was my second-to-last pair of pants? Newsflash: A week earlier I did the exact same thing during group pump training—and at a hydrant along the busiest road in town at midday, with tons of traffic. Facing the rig, opening the hydrant, I felt that familiar pop. "No way," I gasped. Then, when I feel a cool breeze from behind, I realized I'd just given the entire town a free strip show. Bolting to the firehouse, I covered my backside with my gloves.

After this happened for the second time, I explained my wardrobe malfunctions over the past 2 weeks to the chief and got the green light to order new pants—ones that could actually handle a butt the size of the Grand Canyon. I'm seriously reconsidering my career choice, though: Maybe it's time to put these stripper skills to good use since my pants clearly can't contain themselves. So long to my old pants—may they rest in peace!

Key Takeaways

You're going to screw up. You're going to spill drinks, fall down stairs, break garage doors, forget tools, or back a rig into something that doesn't move. Welcome to your first year in the fire service. These aren't just embarrassing stories; they're part of the job. Every mistake you make is a brick in the foundation of experience—as long as you own it, learn from it, and keep showing up better the next time.

(*continued*)

> The firehouse isn't a place for perfection. It's a place for accountability, humility, and growth under pressure. You're going to get laughed at, but that laughter builds bonds. It means that you're part of something bigger than yourself. What matters when you mess up is not to hide it, point fingers, or shut down. Instead, raise your hand and say, "Yeah, that one's on me."
>
> Mistakes handled with integrity earn respect. They show that you care enough to do better. And the reality is, every senior firefighter you look up to has had their own embarrassing moments, too, from which they learned and moved forward.
>
> So stop trying to be perfect. Instead, be honest, be teachable, and give a damn. That's what earns trust. That's what earns responsibility. And that's what turns a rookie into someone worth relying on when it really counts.

Final Thoughts

If there's one thing I've learned through every stumble, spill, dent, and disaster, it's that these moments don't define your worth as a firefighter. What defines you is how you handle them. The fire service has a funny way of humbling you fast and teaching you even faster. You'll never forget the times you felt like a walking blooper reel. But one day, you'll look back and realize that those moments taught you the most about patience, resilience, teamwork, and yourself.

So give yourself grace. Laugh at yourself. Remember, the strongest firefighters aren't the ones who never mess up, but the ones who keep showing up, ready to do better. The stories you star in—in particular, the embarrassing ones no one warned you about—will teach you more than any class ever could. They will shape who you become.

Reflection Prompts

- ✠ What was your most humbling or embarrassing probie moment, and what did it teach you?

- ✠ How did you handle a mistake on the job? Looking back, would you do anything differently?

- ✠ What does extreme ownership mean to you? How will you practice it?

14

HAV a Blast

Large-diameter hose (LDH) is typically 5 inches across, but varies between 4 and 6 inches. (If you experience the luxury of dragging a 2½-inch attack line because your probie pulled that cross lay instead of the 1¾-inch attack line, I hope you have enough personnel.) LDH supplies water to your attack line and connects your apparatus to a hydrant. You use it to supply another pump, which could be tagged to a hydrant feeding water from your truck to either a single or multiple apparatus, depending on the assignment. If you're in a nonhydrant district, you're drafting; your water supply is minimal until the shuttle arrives, so use it sparingly.

April 24, 2024

It's a total madhouse at the station. Call volume is through the roof—easily one of our busiest days of the year.

Every year we hold an internship program to give high school seniors an opportunity to get real-world experience, to see if this might be a job they'd want to do once they graduate. Today I have the pleasure of mentoring the intern, and it's her last day with the station before graduation.

The directors of the internship program dropped by for a final picture with the intern and our deputy. When they insist that I get in this picture, I make a suggestion: "Let me pull the ladder out for the picture." The deputy loves the idea. I leave the ladder running once I pull it out to prevent the battery from going dead, even if just for a short time. Diesels are loud when they're running; you can't hear anything but your own thoughts.

Radio Silence

We take the picture, and then conversation strikes back up. All of a sudden, I see one of my shiftmates fly around the apparatus, glaring at us: "Come on, what are you waiting for? We've got a car fire to go to!"

The deputy and I lock eyes: "Uh-oh."

"Did you hear anything?" I ask.

"Nope," he replies. With all the excitement, I didn't realize that I'd misplaced my radio, which I carry with me always, somewhere in the station. In fact, neither one of us had our radios—and he's the one in charge of me!

The deputy boogies to his car, taking the intern with him. Scrambling to find where I last left my gear—it's scattered everywhere!—I finally gather my turnouts and haphazardly throw on my gear. Then I hear, "You're driving." I hop into the driver's seat of the engine. Somehow the ladder is just leaving even though I took a month of Sundays to get ready.

I don't have a clue where we're going because I didn't hear the call come in—it's pandemonium, and my day's just getting started!

Red lights and sirens blare as we speed down the 3 miles to the call. "Engine 3, tag the hydrant just before the street and tie into the ladder." My officer tells me to slow down. I yell back, "Hydrant's coming up in 500 feet."

"Keep driving," he says. I hear the order, having now received two orders that directly oppose each other. So I do as I'm told in the present, pulling to where he tells me to stop. He gets out and immediately heads to the ladder to grab the K12 saw and pike pole, thinking two steps ahead, to gain access to the vehicle latches once extinguishment happens.

The vehicle fire is too close for comfort to a private residence, with trees nearby. The flames begin extending to the power lines directly above. The officer who made the call to tag the hydrant had done so as a safety precaution, given the potential exposures that could ignite if the heat got intense enough or if the flames made contact.

I get out of the ladder and do my 360° assessment. The hydrant is 500 feet behind the truck. *Come on!* It's old and clearly hasn't been operated in years. When hydrants haven't been used for so long, the steamer valve and outlet caps tend to seize up and don't want to come off the hydrant.

The police department shuts the road down since LDH will be supplied; the line will burst if it gets run over. While chaos breaks around me, I'm thinking over my options. I've got my LDH preconnected to the hydrant assist valve (HAV). A HAV is an adaptor for a hydrant with a valve that controls water flow, essentially allowing more pressure. You can fully open a hydrant once the HAV is connected; no water will come out until you open the lever that allows water flow. It's like a kitchen faucet for a fire hydrant.

This is how the truck is set up, and the LDH with the HAV attached is a heavy son of a gun. I need two hands, one for the HAV and the other to grab a hose fold and go. If the steamer is seized then I will have to go plan B: fully dress the hydrant using the two 3-inch shorties (our custom-length hoses), Morse gates, and Storz connectors. This is how we're equipped in my department.

Ask for Help

Our hydrant bag weighs about 75 pounds—so you need two hands for this bag, no matter how strong you are. Even if you're the strongest member of the department, there's no way you can be efficient while trying to carry the HAV, LDH, and the hydrant bag at the same time. Except for the HAV, the hydrant bag contains everything you need to fully dress a hydrant: a breaker bar, hydrant wrench, additional adaptors, and Storz connections. This gives you plans A (as is), B (with adaptors), and C (with additional connections) for hydrant operations. If all options go to hell, then call for help proactively. Communication is key on the fireground.

Hopping up onto the tailboard, I take my chances, choosing between the hydrant bag and the preconnected HAV. I grab the first fold, then the HAV, before jumping down. Pulling LDH from this engine, which has a sharp boxlike edge that creates excess resistance, is like dragging a weighted sled. It's tough to keep momentum going, and it exhausts you quickly. In addition, the hose sits more than 10 feet off the ground when it comes off the bed. I'm pretty tall, but even I struggled to reach it. I'm pulling and struggling—all that effort only to reach 500 feet.

Here goes nothing. I hope that steamer valve will work and breathe a sigh of relief that it hasn't seized. I communicate my progress to the incident commander: "Hydrant's tagged, call for water when ready, connect the LDH to the intake of the ladder."

The ladder operator relays back: "All set, send water." Water is successfully established. The fire is extinguished and we wrap things up.

Vehicle towed, the road is cleaned up, chaos under control. Incident command clears all units, and we head back to headquarters. Usual routine sets in as we clean hoses, tools, and packs. We keep our personal protective equipment as new as we can, to prevent exposure. As the day winds down, so does the chaos. We had a solid shift, and now it's time to take our 24 hours off and gear up for our second and final tour of the week.

Do the Right Thing When No One Is Watching

April 26, 2024

Next shift, the private who was on the receiving end of that supply line during the car fire pipes up in front of the crew who weren't there for the vehicle fire on the previous tour. Apparently, he had been at the pump on the ladder quint, watching me work the whole time.

"Dude, did you hear about what Nichole did?"

Anxiously, I wonder where this is this going, but then he continues: "This kid hauled 500 feet of LDH uphill over her shoulder, with a HAV in hand, and tagged the hydrant like a boss at that car fire the other day!"

I try to shrug it off as no big deal, but he makes sure I know how impressed he is: "That was savage."

Key Takeaways

Sometimes the job isn't about the glory; it's about the grind no one sees with no one coming to rescue you when it sucks. It's about doing what needs to be done when your radio's missing in action, orders are flying, and chaos is the only certainty. That's the kind of day this was: full-on fireground madness, including me trying not to flatline before even reaching the hydrant.

You never know who's watching, and more important, you should act like no one is. Because doing the right thing when no one's around to clap for it? That's what builds reputation, character, and trust, following you tour after tour. It's nice being called savage every now and then—you just have to earn it.

I didn't haul that hose to be a hero. I didn't dress that hydrant thinking someone would give me a shout-out later. I did it because the job needed to be done. That's the difference between someone who clocks in and someone who shows up. This job rewards quiet consistency more than loud confidence. So keep grinding. Keep showing up. And keep doing the right thing even if your radio's dead, your gear's scattered, and it feels like no one's paying attention. Because someone always is.

Final Thoughts

There's no shortcut for grit on this job. You can't fake your way through the hard parts, and you can't wait around hoping someone else will swoop in to save the day when things get tough. The fireground doesn't care how tired you are, how heavy the hose feels, or how chaotic everything seems. It just asks whether you are ready to handle it.

This chapter is about digging deep, staying steady when everything around you is a mess, and proving to yourself that you can rise to the occasion. Not for the praise, not for the glory—but because it's the right thing to do.

So the next time the job hands you a moment that sucks—when you're sweaty, exhausted, and wondering if anyone even notices—remember this: Those moments are when your reputation is forged. In them, you become someone your crew can count on. And that's what really matters.

Reflection Prompts

- ✠ Think of a time on the job when everything felt chaotic. How did you handle it? What would you do the same or differently next time?

- ✠ When was the last time you did the right thing without expecting recognition? How did that feel?

- ✠ Have you ever faced a situation where no one was watching but you chose to give 110% anyway? What did you learn from that experience?

15

SMALL TOWN THINGS

May 2, 2024, 1852 hours

We're dispatched for a well-being check to a vague location—no exact address, just a general "out there somewhere." We can't seem to find the person in need. Guess who does after 15 minutes of searching? Police. Of course they do, because they're always first on scene. An officer greets my partner and me and gives us the rundown. The man receiving this well-being check is a senior citizen who lives nearby, doesn't want a ride home, and is completely harmless.

A passerby called in this check having seen the person in question walking down the middle of the road. A majority of 9-1-1 calls are not accurate and may be blown way out of proportion. In this case, the old man walking in the street claimed, "I've got the right of way, and traffic can navigate around me"—possibly senile or just plain stubborn.

My partner asks him, "Are you OK? Having a medical emergency? Anything we can do to help?"

The senior citizen, with the energy of a man half his age, replies combatively, "I'm 71 years old, just out walking! I locked myself out of my car, went to get ice cream, and now I'm having a nice walk."

My partner asks "Why don't you walk in the grass off to the side of the road?"

The old man huffs back, "The grass makes me trip, it's uneven, and I can't walk on it. You expect someone my age to walk on that? I'm not doing it!" He's determined to keep walking down the road toward his home, no matter what we say.

"Sir, why not let the cop give you a ride? We don't want you getting hit by a car."

In response, just a stern grumble: "Nope, I'm all set."

Police look at us, and we look back at them, both confused by the situation. We can't force the guy to get a ride, we can't force him to be evaluated, and we can't leave him walking in the middle of the road, where he might be flattened into a speed bump.

At this point, the resident was about 500 feet from our ambulance and the police car. "I'm gonna roll behind them so they don't get hit," I say.

Now the police officer and I hop in our vehicles and start a game of leapfrog down the road. The police car pulls ahead, then the ambulance, back and forth. The old man keeps mumbling, "My house is just right there," but keeps walking, reinforcing our concern for senility.

In our small town, we've got a close-knit relationship with the police department and the local residents. We give the time it takes, even on calls like this. Effort counts, and we take pride in what we do to serve the community with a high standard of care. No one wants this old man to get run over. And the officer's careful not to provoke the man into becoming more disgruntled than he already is. The old guy keeps refusing a ride and won't tell us his house number, so we all just keep strolling down the road.

Finally we arrived at the resident's house, 3 miles and 1 hour later. He plops down on in a rocking chair on the front deck like it was just a nice afternoon stroll and as if we don't even exist to him. Before leaving, we make sure his house is unlocked and ask if he needs anything else. With all the charm of a grumpy king, the man waves us off like peasants with a dismissive "Leave me alone." Police clears the scene, and so do we.

It's a day in the life of a small town. As Forrest Gump reflects, life is "like a box of chocolates. You never know what you're gonna get."

Key Takeaways

Small towns come with their own kind of chaos. It's not always a fully involved structure fire or a dramatic rescue; instead, sometimes it's an hour-long slow roll behind a grumpy old man with an ice cream cone who just wants to get home. Even though they may seem ridiculous on the surface, these moments matter. This job isn't always about lights and sirens. Sometimes it's about patience; other times, it's restraint; and more often than not, it's just showing up and giving a damn, even when you're being shooed from the scene like you've been voted off the island on *Survivor.*

In a small town, you don't just serve a community; you become part of it. Every weird call, awkward walk-along, or sidewalk exchange adds up. You build trust not only with the people you rescue from burning buildings but also with those who just need someone to look out for them on a random Tuesday. These calls might not make the news, but they're where the heart of public service really lives.

Final Thoughts

You can't always control the call. Sometimes it's a medical emergency, a fire alarm, or warring neighbors calling in each other's cooking fires. Other times it's just a stubborn old man who thinks the middle of the road is a perfectly reasonable sidewalk. Regardless, what you can control is how you carry yourself. Do your job with grace, humor, and care, even when it's inconvenient. Especially when it's inconvenient.

The truth is that people remember how you made them feel far more than what you did. How you treat these seemingly small moments shapes your relationship with the entire community. The effort you put into the little things—even leapfrogging down a road alongside a stubborn and possibly senile man—says everything about your character.

A reputation for effort travels fast through word of mouth in a place where everyone knows everyone. So lean into the weird calls. Show up with patience. Keep your cool when the situation gets ridiculous. And always remember, some of the strongest connections you'll ever make won't come from saving lives, they'll come from simply giving your time, even when no one's clapping.

You can't control the event or the outcome, but you can control how you react. For every action, there's an equal and opposite reaction. Do your job, do it well, and treat every call, even the ones that seem like nuisances, as if it were your own emergency.

No day is ever the same in the best job in the world. If you love what you do, you'll never feel like you're working a day in your life. Have gratitude and may peace be with you.

Reflection Prompts

- ✠ What's a day in your life on shift like for you?

- ✠ What's one moment that felt small at the time but later reminded you why this job matters?

16

A SLICE OF BURNOUT PIE

The high of being on the job after finally making it out of the academy can come crashing down on you if you aren't careful. Adjusting to what this job demands, with its capability to help or hurt you, is not easy or comfortable. It's not supposed to be.

Finding balance within the chaos is a necessary challenge, or you may come to an unexpected uh-oh moment in your career. Mistakes will happen—in fact, they're unavoidable—but every mistake can be a lesson if you let it. As Mr. Clegg, my academy instructor, once told me,

> *Your first year is for making all the mistakes you can. Every day and every year means growth. You'll never be perfect, but that's how you learn. Owning up to your screwups, learning from them, and growing, that's what being a good firefighter is all about. That's what makes a good officer. It's part of the process, and it's part of what makes you stronger. Life isn't sunshine and rainbows, and firefighting sure as hell isn't. But you can take the hits and evolve into someone stronger, someone you never imagined you could be.*

How It Began

April 7, 2024

The gold standard in this profession should be to walk in with drive and a smile that can't be erased. Quickly, though, the smile is gone, buried under the weight of exhaustion and despair. Right after the academy, this job felt like a vacation, not like work. Then, after only a few months, suddenly I dreaded showing up for my scheduled 24-hour shift, not wanting to be there a minute longer than needed. Burnout already, 6 months in—dragging myself through each shift, barely recognizing the person I had become.

This began when eligibility for overtime kicked in. Jumping at the chance, I dove headfirst into a hole out of which I needed to be rescued. In 1 month alone, I worked almost 400 hours, 85 to 90 hours a week. The paycheck was intoxicating, with a single week's pay approaching what I previously made in a month. The hustle consumed my identity. Earning the nickname Overtime O'Donnell, I wore it like a badge of honor, while falling deeper into the abyss.

Fixated on helping the department members who had been doing this job longer than I've been alive, I sacrificed myself to give them some relief and time with their families. Then out of nowhere, I hit a wall—hard. You can only do it for so long, until you just can't.

Early on, my shiftmates had tried to warn me about hitting the wall. They pulled me aside, telling me to take it easy, so I wouldn't burn out. Being stubborn as they come, I was driven by the misguided belief that I could handle it.

Burnout crept up on me insidiously—a silent killer, like carbon monoxide poisoning. Overloading my thoughts, overanalyzing every single mistake made, it consumed me. The smallest errors became mountains in my mind, and I started to feel worthless, like a failure.

The Reality of Burnout

May 7, 2024

En route to a structure fire, I was driving the engine, exhausted from 3 days of relentless work. It was raining out, and road conditions weren't ideal. The captain yelled at me, "Slow down, go around the car in front of us!" My brain, already shut down, was unable to process what he just yelled at me. I didn't see the car in front of us, and we began to skid. With hardly an inch of breathing room around this family in their vehicle, I thought it was all over and saw my life flash in front of my eyes. Not a great feeling by any means.

I blacked out, and the next thing I remember was hearing captain Arventos say, "If I wasn't paying attention, you'd be the one telling my wife and kids why I'm not coming home tonight." That made my stomach drop. I had no recollection of the event itself; the car had been in front of us, and then we were going around it.

Somehow, we survived, but I could have killed my crew, myself, and the family in that car. Numb, I drove the rest of the way to the structure fire in a daze, with my heart pounding. There was zero conversation about it that night. We did our job, extinguished the fire, got the job done, cleaned up, and a few hours later, went to bed.

May 9, 2024

Captain Arventos pulled me into his office, looking me dead in the eyes. "If I hadn't been paying attention," he said quietly, "we'd be dead." He said it was like I wasn't even there, like I'd been on autopilot when he was shouting at me, trying to get my attention. He asked me what was going on in my brain that night, and I felt the weight of what nearly happened. I had screwed up badly and knew it. This was a "Dear Chief" moment.

That night played over and over in my head, an endless loop of shame and regret. All I could think about was how close I'd come to a nightmare headline: "Local firefighter kills crew driving recklessly." I had my commercial driver's license, along with training and prior experience driving big rigs, yet I still let my crew down in a way that felt unforgivable. The conversation wasn't just a wake-up call, but a deep, twisting pain that wouldn't go away—like an itch you can't scratch. I'd betrayed the trust my captain had put in me to keep everyone safe.

I dreaded that I'd lost everything I'd worked so hard for and had failed the individuals who counted on me the most. My dream job, for which I'd sacrificed everything, was slipping through my fingers because I couldn't keep myself in check. Forcing myself to realize what I'd been ignoring for too long, I accepted that I was burnt out and had lost touch with not just myself but reality. Pushing so hard, I'd lost all sense of who I was, what I was doing, or why any of it mattered.

Obsessed with being the perfect probie blinded me to risks, outweighing my decision-making capabilities. Consequently, I became the reason for the near miss. Facing my mistake was the hardest lesson, but I had to accept the reality of it, learn from it, and move on.

Taking Care of Yourself

Isolating, not eating, and completely neglecting workouts, I shut out everyone who mattered to me, including my best friend. I didn't have a clue as to what was happening inside my head, but I knew I felt horrible. The people who care about me had no idea what was going on and started worrying that they were the problem.

Too selfish to let them in and too wrapped up in my own mess to see how much I was hurting them, I lashed out and the people I love took the hit—simply because I didn't want to admit that I was struggling. Don't make the mistake of shutting out those who love you.

With the passion and drive I had for this work, it was nearly impossible to tear myself away, even though I knew I was in trouble. Still, time away from the job was the only way to redirect myself and get back to where I needed to be. I had to remember what it felt like to live. Once I took time off, I finally began to heal, and it was like the clouds parted. The constant stress and overthinking eased up, and I started focusing on things that mattered to me, things outside of the job, and it changed everything. I hadn't felt so relieved or happy in months. This break saved me.

Still, the scars are there. The shame of almost ending everything in a moment of exhaustion will never fully go away. It's a reminder that I'm not invincible—no one is. Burnout is real, and it will destroy you if you let it. So take the time you need. Don't make the same mistake I did.

Recommendations

Don't be like me, running headfirst into burnout. Don't ignore the warning signs. Address issues before they drag you down. According to recent research,[1]

> *Several factors . . . influence firefighter suicide, including the effects of post-traumatic stress, sleep problems and substance abuse. The potential effect of post-traumatic stress in regard to suicide was recognized for years for police, military veterans, ambulance crews and disaster workers, as are other factors, such as burnout and meaning in life. Burnout for those who work in jobs that help others is defined as the experience of emotional*

1. Matt Coffey, "Health and Wellness: Firefighter Suicide and the Link to Burnout and Lack of Meaning," *Firehouse.com*, April 17, 2024, https://www.firehouse.com/safety-health/article/53098148/firefighter-suicide-and-the-link-to-burnout-and-lack-of-meaning.

exhaustion, loss of empathy and loss of a sense of personal achievement. Burnout is well known to be linked to depression and suicide in nurses and doctors. Although burnout is not well studied in firefighters, some research from Poland shows that it's a significant factor in Polish firefighters' mental health problems.

Key Takeaways

Burnout can sneak up fast, even when you love the job. If you don't set boundaries, the passion that drives you can also be what destroys you.

Overworking to help others, though noble, is dangerous if it comes at the cost of your own health and safety. Your desire to support your crew should never overshadow your responsibility to stay sharp and safe.

Mistakes are inevitable, especially in your first year. What matters is owning them and learning from them, so that you grow into a better firefighter and leader.

Burnout affects others beyond yourself. It strains your relationships, erodes your mental health, and puts others at risk, both at work and at home.

Listen when people warn you. Fellow firefighters, officers, or friends who see you struggling aren't trying to slow you down; they're trying to save you.

The job can't be your whole identity. If you lose yourself in the hustle, you risk forgetting what it means to truly live.

Stepping back isn't weakness; it's survival. Time off, rest, and reconnecting with life outside the firehouse are critical to staying in the fight for the long haul.

(continued)

> Mental health matters. Burnout is real, and left unchecked, it can lead to serious consequences, including tragedy on the job or even suicide. Don't ignore the warning signs.
>
> You can recover. Healing is possible with humility, self-awareness, and a willingness to prioritize your well-being.
>
> Remember your *why*. When the job weighs heavily on you, pause, breathe, and remind yourself of the reason you chose this career in the first place.

Final Thoughts

Everything happens for a reason, and you can learn from every mistake. Take care of yourself; don't let the job consume you. Your mental health matters—and more importantly, you matter. No one—officer, chief, or shiftmate—has the right to make you feel guilty for taking care of yourself.

Burnout is real and will destroy you if you let it. It could even kill you. You will hit a point where your body and your mind just can't take it anymore, and it will make you physically sick. Mental health isn't something you can brush off; it's critical, so take time when you need it.

You've earned your time off, and you need to take it for your own sake. Take this job seriously, but don't dig yourself into a hole you can't climb out of. When you become overwhelmed, take a step back, breathe, and look at the bigger picture; most of the time, it's not as bad as it seems. Sometimes the only way to move forward is to swallow your pride, refuel, and remember why you took the oath in the first place.

Reflection Prompts

- ✠ What's the biggest obstacle you've ever faced?

- ✠ How did you overcome it, and what lessons did you learn?

17

SMALL THINGS MATTER

June 19, 2024

In the fire service, knowledge is your lifeline. Success not only entails being good on the nozzle but also requires knowing the inner workings of your department, including its structure, operations, policies, and history. Because when things go sideways, the most trusted person on the scene isn't always the most experienced, but the one who's best prepared. Especially in a small-town department, knowledge becomes survival.

We're not a big-city crew with three firefighters on every rig. In our department, a crew may consist of just the captain and the driver when both ambulances are out. We have only four captains (one for each shift, no lieutenants), a fire prevention captain, a deputy chief, a chief, and 20 privates.

On the fireground, the captain does the 360° assessment while you pull the attack line and work the nozzle. You're also charging the line, dead-manning the pump, and getting ready to work. (Dead-manning the pump basically means putting the truck in pump, charging a hose line, and then going to work without having anyone as the pump operator while hoping nothing catastrophic happens, like losing your water source.)

If a trapped victim is reported, you're searching with your captain, not waiting. Two people are acting as a multiple-company operation. Mutual aid? Could be 10 minutes away, or could be 30 or more.

We don't have specialties; we are the specialty. One minute you're the interior attack line. The next, you're extrication, medical, ventilation, or drafting from a pond in the middle of nowhere. If you can't be flexible, then you're a liability. You've got to know your equipment, your truck, your town, and your expectations. And most importantly, you've got to know your strengths and weaknesses.

August 25, 2024

Night shift means minimal staffing. I was alone at headquarters, in the middle of my workout, when the captain radioed me. It took a full 5 minutes for me to realize he was even calling. "Tone out the callback," he ordered. I froze. I had no clue how to do it. Swallowing my pride, I called dispatch for help. They bailed me out and handled the callback.

When the captain got back, he reminded me that this was covered in training. In that moment I was irritated with myself. If I couldn't be trusted with something as simple as a callback tone, how could I be trusted on a structure fire? What if I had to grab a truck, tie into an engine, or run the pump alone? Trust is everything in this job. And I'd just lost a bit of it so early in my probation. Again, though, when inevitable mistakes occur, learn from them and move on. It's not the end of the world.

Know Your Department, Know Your Town

August 30, 2024

Later I learned I wasn't even supposed to be left alone that night. That wasn't a probie mistake; it was a chain-of-command issue. Still, it reminded me that not knowing something can be costly, even if it's not your fault.

That's when I realized that the job isn't just tactics. It's knowing your department, your people, your policies, and your town. Our population is 65% senior citizens. Nearly 12% are veterans, including some veterans of World War II and Vietnam. We operate as a fire-based EMS system at a strict ALS level. When we're not fighting fire, we're delivering critical care. Hence, the trust of our community is everything.

If a community member comes to the firehouse, asks for directions, and you don't even know the road they're talking about, what message are you sending? You're a public servant. You swore an oath. And you don't even know your own town? Being proactive isn't just about hose loads or running drills. It's about learning the place and the people you serve.

The community takes pride in knowing where their tax dollars are going. If we make them feel indifferent or unwelcomed, we risk losing their support. And in that case, you can kiss your grants and funding goodbye.

Making an Impact

This job is not just knocking down fire. It's staying late to relearn a procedure you forgot. It's running your own drill because you haven't tied a knot in a month. It's reading the policy manual on a Sunday morning while others are scrolling on their phones. That quiet, behind-the-scenes effort makes you reliable. And reliable firefighters? They're the ones officers remember when it's time to delegate important tasks.

People often treat department and community knowledge as optional. But that's wrong. As a firefighter, you serve people, and you can't serve people if you don't understand them. Know the layout and the hazards; know the population and the stories. Every piece of information helps you to make better decisions on the fireground and the community to see you as someone who cares about them.

My small town is the home of Friendly's Ice Cream, which started in Springfield, MA, but moved their headquarters here. I lead with this detail when someone asks where I work. It opens the door. People smile and get curious—and boom, you've made a connection!

The same thing happens when people ask what I do. My reply: "I have the best job in the world. I'm welcomed into homes on the worst day of someone's life. I give them comfort, closure, and compassion, and I try to make their bad day better." That always gets people talking. That's where trust starts.

It's incredibly rewarding to see the reactions on the faces of local community members when we pull up anywhere in town. The deeply positive relationship we have with the public is honestly one of the coolest experiences ever. People don't normally get excited when random strangers show up—but when it's the fire department, an instant sigh of relief and usually a welcoming smile greet us when we arrive.

Small gestures matter. When you give a damn, the community doesn't take it lightly. The result is a unique human interaction.

Key Takeaways

Knowledge in the fire service is about more than competence; it's about dependability. When things go wrong, people naturally look to the ones they trust to handle it without question. That trust isn't given because you wear a uniform. It's earned when your crew knows that if you hand something off to them, it's been handled. That level of trust comes from being prepared, by taking the initiative to know your job, your town, and the systems that make your department run. That makes the difference between being a probie people constantly have to babysit and being one they won't overlook.

Anyone can show up to a drill, crush a task, and feel like they've got it. But readiness isn't proven in a controlled environment. It's proven when you're tired, alone, or under pressure, and you still get it right. Being consistent isn't glamorous. It means revisiting the boring stuff, going over tones, maps, and department standard operating procedures, even after a long shift. It means asking questions when no one else is watching. You'll never feel fully ready, but being prepared anyway is what builds trust.

(*continued*)

> The moment you stop being teachable, you stop being useful. Every shift is an opportunity to get a little sharper, to notice something new, and to ask one more question that might matter on your worst day. Stay curious. Stay hungry. The firefighter who keeps learning is the one who's always ready.
>
> You want to stand out? Don't be the loudest. Be the one who knows their job so well that no one has to second guess them or their capabilities. That's how to build trust. That's how to build a legacy. That's how to lead, long before you earn rank.

Final Thoughts

You can be a great firefighter on paper, but if you're not dependable in real life, the paper doesn't matter. The ones who stand out care enough to learn it all—not just on the fireground but behind the scenes, including administration. This job doesn't require you to be perfect, but it does demand that you be prepared. This job humbles you. It exposes your blind spots and forces you to grow.

Once you know the job, know your department, and know your community, you become someone people can count on. That's what makes an exceptional firefighter others can learn from. Your name might not be remembered, but your impact will.

Reflection Prompts

- What local fact or department tradition could you learn today that would build community trust?

- What's something small that you've avoided learning because it didn't feel urgent?

18

BE A FIREFIGHTER, NOT A WOMAN IN THE FIRE SERVICE

November 15, 2019

Nineteen years old, I was working on a tree crew. Green as they come, I got hit with the "You're female, you don't belong here" attitude right away. Because I was inexperienced, I spent my time dragging brush and learning while having the "privilege" of working with a foreperson who could have given a master class in poor leadership. I was the only woman employed by the company within 500 miles, and eventually became the youngest employee promoted to foreperson in my division, which covered multiple states.

My supervisor was the son of the former president of the company, so he had license to do whatever he wanted with no repercussions. His favorite game was cutting brush, driving the truck 500 feet past the mess he made, and leaving me to drag everything to the chipper. He'd do the easy bucket work and then park his butt in the driver's seat, playing on his phone while I broke my back cleaning up his mess.

When you're a foreperson, you work with your crew, not against them. But he didn't care, helping only when management was around. When it was just the two of us, he'd be useless. His goal was to make me quit, but since I wasn't raised to give up, I worked through it, no matter how ridiculous it got. He treated me like I was stupid, talking down to me. For example, saying, "You're a girl. You don't even know how to start a chain saw."

It was beyond infuriating. Later, I found out he was telling management not to assign me to anyone else, even if I asked. My diligence and work ethic made it so he didn't have to do anything when I was around. He purposely wasn't signing off on my chain-saw credentials, stating "She's not ready," so that he could do the cutting while I did the grunt work of dragging. He reported to management that I didn't know how to run a chain saw, leaving out that he never took the time to teach me.

After 7 long months, I wasn't going let myself be held back any longer. Once I was finally freed from his truck, my career truly began to take off. I had felt stuck for so long, like I was failing before I even had a fair chance. But the moment I moved to a different crew, everything changed. My confidence grew, my skills sharpened, and within 15 months, I earned the rank of foreperson.

That experience, as painful and frustrating as it was, taught me one of the most valuable lessons of my career: Never forget how it feels to be overlooked, dismissed, or held back. I promised myself I would never do that to someone else. I'd be the kind of leader who lifts people up, not one who keeps them down—because I know how much it hurts, and no one deserves to feel that way.

Call 9-1-1

After I was promoted, things didn't magically get easier. You thought having a lazy foreperson sounded irritating? Now I had a groundman who was upset that I got the promotion and he didn't and he nearly killed me because of it.

May 5, 2021

One day, we were working in the small town of Granville, MA, trimming around utility wires. I was in the bucket 50 feet up, trimming trees over the main power lines. Out of nowhere, smoke began to billow from below my bucket. I was hovering over live wires carrying 27,000 kilovolts—a voltage so powerful that if I were to come within 3 feet of the utility wires I would be burnt to a crisp (hence the bucket that's grounded but that doesn't stop the path of least resistance if you aren't grounded properly)—with my truck on fire! As I tried to manipulate the hydraulic pistol-grip controls to rapidly lower the bucket as close to the ground as possible, all of my hydraulic lines burst in midair. Now I was stuck 12 feet up, with hydraulic fluid everywhere.

"Shut the truck off, kill the selector, and call 9-1-1!" I yelled to my groundman. There was absolutely no cell service, though. My bucket was smoking, there was an active flame, my fire extinguisher was a dud, and my groundman was cutting brush to make himself look busy instead of helping. I had to take action quickly, and get down to safety.

Taking a leap of faith, I jumped out of the bucket, landing hard on the cage 12 feet below with a grunt. Adrenaline padded my fall, and any pain was better than burning alive in the bucket or getting fried by the wires.

I needed to call the fire department (and then my boss, although he could wait). I walked 2 miles to find cell service. By the time I found enough service to make a call and walked back, the truck had extinguished itself. When the mechanic finally showed up, he took one look and said, "Someone overtightened the bolts on the rotation joint with a torque wrench. That's what caused the hydraulic lines to rupture." It surely wasn't an accident.

That's when it all clicked. Every day, after I locked up my truck, this one groundman stayed behind. There were no cameras in the parking lot where we kept our bucket trucks, so he had free range. He had tampered with my equipment, knowing exactly what it could lead to. This person had a criminal record, having served 5 years in prison, and acted like he had nothing left to lose.

In hindsight, a pattern of sabotage was evident, as if he was trying to take me out of the game once and for all. He'd offer to prep my chain saw before I went up in the bucket, which in hindsight was a massive red flag, and I'd noticed that every time I worked with him my chain-saw bolts were loose, almost as if he wanted the chain to fly off midcut. Loose nuts and chain saws are a lethal mix! I caught on, though, and started checking my saw right before I cut. Ultimately, the fear that if I didn't check my equipment, either myself or someone else would get seriously injured or worse, trained me to always run through my equipment before use, and I brought this mentality with me into the fire service.

Check Your Equipment

If you don't check your pack during your truck check and it fails you as you conduct a vent-enter-isolate-search in a building, you will become a hazard and have failed the person you're trying to rescue. What about your saw on the truck? If you didn't start it or check the fuel, what happens when you need it, but it won't start, or when you run out of fuel halfway through ventilation? There isn't a good ending to this scenario, so be proactive and check your equipment. Learn from my experience that even when someone says they have already checked it out and says that you're good, you must still thoroughly check your equipment yourself, because even someone with the best intentions can still miss something. And if someone who means well can fail you, what happens if there's someone who doesn't?

The Reality of Discrimination

That experience taught me early that discrimination exists. I faced it head-on, but fortunately, I haven't had that kind of experience in the fire service. I've been treated with respect because I came in ready to work, prove myself, and focus on the job, not my sex.

That's not to say discrimination doesn't exist in our field. It absolutely does. There are still people who will judge you before you even pick up a tool. There are people who will doubt you—not because you can't do the job, but because of their own bias, insecurity, or ego. You won't change everyone's minds. But you can control how you show up and prove them wrong through actions, not words.

How You Show up Matters

If you want to overcome negativity or bias, the first step is to check your attitude. Are you there to work, or are you looking for special treatment? Because in this job, nobody cares that you're a woman. They care if you can pull your weight. A stigma to being female in the fire service exists only if you let it.

The best way to beat discrimination? Outwork everyone. Be so solid, so reliable, that nobody can question why you're there. Don't play the woman card. Don't expect praise for doing the job you signed up for. Don't expect anyone to hand you anything because of your sex. This is a team game, and if you're part of the team, you've got to earn your place like anyone else.

The Right Kind of Reputation

June 2024

I've had the privilege of working alongside incredible women who've truly earned their place in the fire service. One of them is my friend Brittany Doanne, a firefighter and paramedic in Massachusetts who was recently promoted to captain. She's earned every ounce of the respect she commands—not just because of her rank, but because she puts in the work even when no one's watching. Another is my friend Jennifer Bratco, a firefighter and paramedic who also serves as an aircraft mechanic and 1LT in the United

States Army. Walk into a room with either of these women and your first thought is, "Yeah, I want to go into a fire with you." And if you've ever actually been on a call with them, you know they've got your back before you even ask.

They've built their reputations the right way, through hard work and determination, without succumbing to a "pick me" or entitled attitude. I would trust them with my life, no questions asked. They're the ones who show up, get their hands dirty, and prove every day that they're here for the right reasons.

Don't Become a Statistic

I've seen the other side of things, where women think they can play the woman card, by taking inappropriate shortcuts to get ahead. Some walk into a room like they're the queen bee, and you can feel the energy shift instantly. This job is hard enough as it is, so we, as women, should be finding ways to work together building one another up, even when we don't see eye to eye.

In this field, your reputation is based on your work ethic, never your sex. We're a team, and if you can't handle working with someone because you feel threatened or aren't the center of attention, then you should take a long look in the mirror and ask yourself what you're really here for.

Lead by example—I can't emphasize that enough. If your focus is always on being better than others or you see everyone as competition, that mentality will get you nowhere, and you're going to rub people the wrong way! One day that woman could outrank you and become your officer. How you treat her now is something she'll never forget, so don't view your as a threat. Instead, think of her as your little sister, as someone who looks up to you. After all, this is a brotherhood and sisterhood, and we all gel into a family. Always do a mirror check on the way you treat others, and correct negative attitude (which social media has a way of reinforcing) before it bites you in the butt.

Key Takeaways

This job isn't about sex; it's about work ethic. What matters is how hard you're willing to work, how you show up for your team, and whether you can be counted on when it matters most.

People will test you. Sometimes because you're new. Sometimes because they're insecure. Sometimes because they're just not nice. Don't take it personally; instead, let your actions prove what you're made of.

Check your equipment every single time. Your safety—and the safety of your crew—starts with you. Never assume someone else got it right, regardless of whether they meant well.

Step up. Take on the tough tasks. The fastest way to earn respect is to be the person who's not afraid to do the hard work.

Don't take shortcuts or expect special treatment. Don't rely on anything other than your work ethic to prove you belong.

Build your reputation on integrity, consistency, and teamwork. What people will remember is how you worked, not who you are on paper.

How you treat others, especially those newer than you, matters. Set the bar high. Be the person who builds others up, not the one who tears them down.

Final Thoughts

Not everyone you work with will have your back—that's the reality in any field. Some people will challenge you because of your background, your confidence, or simply because they can. But you can always control how you respond.

If you're here to work hard, earn your place, and put the team first, you'll thrive. Every tough shift, hard task, or person who tries to knock you down gives you a chance to prove yourself.

At the end of the day, what matters isn't your sex, background, or title, but whether you're reliable, capable, and someone your crew trusts when it counts. Be that person. Be the one people want beside them on their toughest day.

Rule number 1 for women is don't play the woman card. Shut up, do the work, and prove that you belong here, just like everyone else. This is a sex-neutral job. No one cares that you're a woman in the fire service; what they do care about is if you're here to work and if you can do the job efficiently without having to be babysat.

Reflection Prompts

- Have you ever noticed or experienced situations where someone was treated differently because of their gender? How was it handled?

- What steps does your department take to promote fairness and respect in the workplace?

19

WORK/LIFE BALANCE

August 12, 2024

After almost a year on the job, I'm drained. This grind, doing the same thing over and over, has taken a toll. I used to think that work/life balance meant sticking to my shifts, hitting the gym, eating clean, and keeping myself in check. But that's not balance.

Balance is about stepping outside the routine and actually living. It's not just about the daily grind, it's about finding moments that remind you why life is worth living. For some, those moments might be wild nights at the bar, but for me, it's finding time to do the things that remind me who I am. Not just what I do for a living but hitting the trails, traveling to new places, and chasing that next adrenaline rush.

Last week, I hit the wall. My body was screaming at me, "You've been pushing too hard!" It wasn't just exhaustion. No, it was more than that: It was burnout, which creeps up on you and makes every task feel like climbing a mountain.

I was a mess, sluggish, and unmotivated, feeling like my own fire was extinguished. It wasn't one of those minor slumps that a nap or a cup of coffee will fix. It was the feeling that I couldn't do this anymore. I knew from experience that if I didn't get away from work, I was going to end up either breaking something or breaking down. Hitting burnout early in my career taught me a lesson: Take the time for me when I recognize the signs of imminent burnout. This time, my solution was taking off my birthday week to reset by doing things to remind myself that I'm more than just my job.

I'm not someone who usually takes time off. I don't like sitting still, and the idea of self-care has always sounded like nonsense to me. But at that point, about to turn 24 years old, I had realized that burnout was an abyss, and I needed to step back from it before it swallowed me whole. I unplugged, shut off all work notifications, muted the scanner, and logged out of the callback application. No overtime, no work. Just 7 days of personal time. I told myself to actually have some fun for once. The job and the pay would still be there when I got back.

Into the Dirt

August 15, 2024

I kicked off the week hanging out with my best friend Alycia's kids: Tommy, Vincenzo, and Gianna. Tommy and I went riding motorbikes. The second I got to the motocross track, surrounded by the sound of revving engines and the smell of dirt, all my stress, all the nonsense that had been piling up was gone. The next day I took Gianna to an indoor trampoline park. If you want to feel like a kid again, go to a trampoline park and lose all sense of adult dignity. We bounced around without a care in the world, climbed rock walls like we were in an action movie, and blew more money at the arcade than I care to admit. Worth every penny: No stress—just pure, ridiculous joy.

Thursday it was back to motocross life with Tommy. Even though I hadn't ridden in a while, after a few laps in, it felt like I'd never taken a break. With that engine rumbling underneath me as I tore down the track, kicking up dust, I wasn't anything but me. I wasn't thinking about work, deadlines, or responsibilities. I wasn't Firefighter Nichole or Emergency Medical Technician Nichole; I was just Nichole. It reminded me of why I fell in love with this sport in the first place: The simplest things can bring you the most joy.

After shaking off the dust, we headed to Cracker Barrel. Somewhere between the chicken-fried steak and the biscuits, I got this wild idea: Why not skydive tomorrow? I mean, I'd already spent the week doing everything I loved, so why not go all in? Life's too short to play it safe, so I signed up to jump out of an airplane at Ellington Airport the very next day. Even though I didn't take Vinny out yet, don't worry; we're going to get some ice cream and go fishing soon to make it up!

Into the Sky

August 16, 2024, 1400 hours

On arrival at Ellington Airport—ready to jump out of a perfectly good airplane—I waited 2 hours, getting hyped with every passing minute. Finally it was time to gear up and jump. I was suited up, strapped into the tandem harness, and did a prejump interview on video.

The ascent through the clouds to 10,000 feet felt surreal, like something out of a dream. The door opened, and it was time. As I dangled my feet out of the plane, everything else in the world disappeared—no stress, worries, or anxiety.

Three, two, one, jump! The free fall, plummeting at 100 miles per hour, was pure, unfiltered ecstasy (fig. 19–1). No fear, just laughter and an uncontrollable grin when you realize you're as free as you can possibly be. It felt like just me and the sky, even with my tandem instructor there. As I was free-falling through the clouds, a thought crept in: Life isn't always about the rush. Sometimes it's about holding on to what matters most, to the people we love, because you never know when they'll be gone.

My grandmother had on her bucket list to parachute and to drive a race car. I kept her legend alive by completing this bucket-list item.

As the parachute opened, I accepted the only downside, which was the harness digging into me. But who cares!

FIG. 19–1. Free-falling at 100 miles per hour

Skydiving was worth every second of that rush, even with a slightly uncomfortable landing.

The Spartan Race Experience

August 17, 2024

Until yesterday afternoon, I had forgotten that I signed up for a 4-hour Hurricane Heat Spartan Race starting at 2:00 a.m. I barely had time to grab a quick shower, pack my gear, and head out by midnight for the race.

Naturally things didn't go as planned, and I got pulled over driving to the Spartan event. The officer's reaction to a 2:00 a.m. race, though, was priceless: "What the heck, a race at this hour?" I showed him the email confirmation, and with a chuckle, he let me go. Great start, right?

Arriving an hour before the race, slightly lost but ready to go, I hiked a mile uphill to the registration booth through gravel and mud, carrying 50 pounds of gear, with a 5-gallon bucket in one hand and regret in the other.

I met my team, consisting of 42 total strangers. It's amazing how fast you bond when you're covered in mud, cow dung, and sweat. By 2:30 a.m., we were warming up, and by 3:00 a.m., we were racing. The first task was to move three cords of wood as a team, piece by piece, a quarter mile down the trail. I grabbed a piece of plywood, thinking, *Let's work smarter, not harder.* We were given 25 minutes to finish, and we finished with 6 minutes to spare. I was relieved our plan worked, because otherwise, we'd have been punished with 100 synchronized burpees. No thanks!

Riddles, obstacles, you name it. We bear-crawled through corn stalks, crab-walked through 4-foot-tall grass—full of ticks and who knows what else—and tied our shoes together with a partner's to crawl through 3 feet of mud in pitch-black darkness.

All together, it was 4.5 hours of exhausting madness. Nonstop work the whole time, but we killed it. We worked as a team, total strangers turning into an unstoppable force. By 7:30 a.m., just as the rest of the world was waking up, we crossed that finish line. It dawned on me that I'd just completed an endurance race despite having had no sleep for over 30 hours.

I had starred in my very own live-action movie over the past 24 hours—skydiving, racing, and feeling more alive than ever. The race tested me mentally and physically in ways I hadn't been tested before. In just a few hours, I had mustered more fortitude and resolve than I'd gotten from some of my toughest life experiences. The ride home, though, was rough. I had an hour drive and was running on fumes. The rest of that day was a blur of campfires, s'mores, and just feeling like a kid again.

Happy Birthday to Me

August 18, 2024

I kicked off my birthday with a breakfast fit for a queen: sausage, eggs, protein waffles, bacon, and coffee! And I treated myself to a five-star Italian dinner that night. I spent 3 hours getting dolled up for my solo dinner, took a million selfies, and headed out. It was the best decision I could've made for myself. My best friend had a beautiful cannoli cake waiting for me, and the meal was pure bliss. Even if dining alone felt awkward at first, eventually it hit me: I'm doing this for me! I savored every bite of the glorious meal, and by the end of the night, I had become friends with the three servers who were taking care of me.

Seven straight days of carefree joy—no workouts, no routines—was exactly what I needed. Taking care of yourself isn't just about sticking to schedules and saving money. Sometimes, it's about stepping out of your comfort zone and doing what genuinely makes you happy. After this week, I felt like a new person.

Mental health matters. Don't be afraid to take time for yourself when you need it. Then, when you're ready, you can come back stronger than ever.

It Gets Harder Before It Gets Better

August 20, 2024

After all that time off, I was completely refreshed and ready to slide back into the grind. But life had other plans. On my first day back, the shift was going great: We were busy, with nonstop call volume, and I was working with one of my favorite partners. The day flew by, with that rhythm you get only with a good partner.

As we sat down for dinner, though, the mood shifted. I got the call: My grandmother had passed away. It wasn't unexpected, but it still hit hard. I felt a strange mix of sadness and relief.

My grandmother had been in decline ever since my grandfather passed. Her quality of life had faded after 4 years in a nursing home and a stroke a year ago. At 84 years of age, she was the sweetest lady you'd ever meet and let me know how proud she was of what I had accomplished (fig. 19–2). As much as it hurt, I knew she was finally at peace, with my grandfather now, probably giving him hell for dying first. Death's an odd thing, both beautiful and ugly, but inevitable in the course of life: When it's time, it's time.

I love you, Meremere. I'll take the reins from here. It's my job to be the strong one now.

Later that evening

My crew told me to take the night off and use bereavement leave. I didn't argue, because they've got more experience with this stuff than me. But I had just taken 7 days off—and now another 7? I wanted to do my job, the thing I love, but I was stuck with the weight of my grief, and my head was not in the right place to work. So I needed this time off more than I thought.

Life sucks sometimes. People say that everything happens for a reason, but when it comes to death, that feels like a load of crap. Maybe this is a blessing in disguise, though; with 2 weeks left until I started paramedic school, this gave

FIG. 19–2. Meremere and me at my Massachusetts Firefighting Academy graduation

me time to study and to focus—to get my head right for the next 2-year journey ahead of me. I've also got big plans for future books on self-help and guidance through lessons learned.

We all face hardships, and we learn through the steps we take to deal with them. What we do in response builds us into who we are. I'm excited to see how I grow as a learner and as a person.

Key Takeaways

Take the time you need.

Mental health matters.

If you see something, say something.

If something feels off, it probably is. Listen to your intuition.

Take care of yourself.

Never take loved ones for granted. You never know when their last day will come.

Go easy on yourself. You deserve a pat on the back for everything you do.

You matter.

Final Thoughts

This chapter of my life taught me about balance, resilience, and what it really means to take care of myself. I used to think work/life balance meant sticking to the grind, showing up for shifts, hitting the gym, eating clean, and pushing forward. But I've learned that true balance comes when you allow yourself to live, step away from the routine, reconnect with who you are outside the job, and fill your cup back up.

Burnout doesn't always announce itself with sirens. Sometimes it creeps in quietly, until one day you realize your fire's gone out. I learned to listen to the warning signs, to reset before it's too late, and to find joy in simple moments—riding a dirt bike, skydiving, and sharing a meal with people who make you feel human again.

We're not only what we do for a living but also who we are when no one's watching. This is reflected in how we care for ourselves and others when it matters most. Life will throw you curveballs—grief, professional setbacks, and other unexpected challenges. But how we choose to respond to these challenges, and how we choose to honor ourselves and those we love is what defines us.

Take the time you need. Protect your mental health like it's your most valuable possession—because it is. Never forget that even though this job is part of you, it isn't all of you. Live a life you're proud of, on and off the clock.

Reflection Prompts

- ✠ What are the warning signs when you're becoming stressed?

- ✠ What do you do to relieve your stress?

20

DON'T FORGET TO HAVE FUN

Ever have the thought about a favorite lifelong hobby or sport that if you get hurt, you won't be able to do it anymore? This starts creeping in more and more as you get older. When you've been doing something for as long as you can remember, it becomes second nature. However, the better you get, the more complacent you become, and complacency can lead to serious injuries—or fatalities! (For more on complacency, see chapter 29.)

When you've got adult responsibilities, serious injury is a risk you can't afford to take. I get it as much as anyone. Once you finally land your dream job, the thought starts to weigh on you: You might have to give up everything you've worked for, so you don't get hurt.

I grew up racing—motorcycle hill climbs, motocross, ice racing, off-road side-by-side, snowmobiles, and the like. When you've been doing something for so long, people get to know you for it. When you're fast, competitive, and one of the best in your area, things get dangerous. In motorcycle racing, for example, if something goes wrong at high speeds, there's no controlling it. The consequences of a crash range from bruises to severe injury—potentially even permanent disability or fatality.

August 7, 2023

My last crash shook me up pretty bad and changed my perspective. The doctors thought I had a brain bleed, but luckily I didn't. After this brain injury nearly killed me, I was scared, and thought I was done riding—I swore I'd never get back on a bike.

It happened as I was racing a motorcycle hillclimb in Monson, MA, 2 months before I got hired on to my career department. I began the fast journey up the mountain on my motorcycle, and next thing I knew, my rear wheel caught the lip of the jump, sending my bike and myself into a full-blown front flip.

The crash knocked me temporarily unconscious. When I got back up, I refused to be evaluated by EMS. Days later, I still had a nasty headache and memory issues. I also began to stutter and had trouble finding words when speaking. Almost 3 years later, these issues have improved substantially, but haven't fully gone away. It still bothers me, but I've adapted. I can't control the outcome, but I can control how I react to it.

September 2023

I received the full video of my accident, and replayed it in slow-motion. My eyes couldn't believe what I saw: My head hit the ground at high speed, and my neck snapped backward. I should have been paralyzed from the neck down on impact. I had a guardian angel looking out for me on that day!

This opened my eyes to how lucky we are and how your life can change in an instant. Don't take anything for granted, because one day your whole life could change. I put riding on hold, putting my responsibility for my career and myself first. I was no longer a kid, with adulthood in front of me.

Recovery

I had a lot of goals when I raced. The main goal was to earn my AMA Pro Hillclimb license, and I came close. The brain injury forced me to tap out, though, after two decades of dedicating everything I had to the sport.

I felt slower at everything. Having a hard time processing words when trying to speak, I was slower to respond. I also had a difficulty understanding information presented in and out of the classroom setting.

Reflection

Slowly I accepted my fate of no longer racing a motorcycle after doing it my entire life. It's bittersweet, but my racing days have come to an end. For a long time, I was unhappy about it, but I've acknowledged the need to move on and learned from the experience.

Racing will always be a part of me, just not at the same level. I loved every moment of it. Even though I'd rather be a lifelong rider than a spectator, I had to slow it down.

If you get hurt, you can't work. Especially during your probationary year, you must show you're responsible and not a liability. If you're irresponsible during your first year, how can your department trust you with any responsibility?

Take it easy, but don't sacrifice all the fun for your livelihood. As in everything else, there needs to be a balance.

Enjoy Life Outside the Job

Eventually I listened to this very advice, from my best friend. I was convinced I'd never get back on a bike after my head injury, but she helped me to realize I didn't have to quit cold turkey. During my probationary year, I took it slow. I still enjoyed motorcycles, but at a less competitive level.

You can't entirely give up something that's been part of your life forever, because it's who you are. If you quit everything you once loved, you'll never be happy. You've got to adjust to find a balance between the hobbies you love and your career.

Don't let your passion or hobby be the reason you lose your dream job. Find the middle ground, so that you don't deprive yourself of something you love. Life isn't always so serious. Go have fun and enjoy what you love—just take it easy.

Key Takeaways

Don't let your job consume your identity. You are more than your badge or title. Hold onto the hobbies and passions that bring you joy, because they're part of who you are.

Balance is key. You don't have to give up what you love, but you do need to approach your passions responsibly, especially when your career and safety are on the line.

Adulthood means managing risk, not eliminating fun. It's about finding the middle ground where you can still enjoy life without jeopardizing your health or your job.

Complacency can be dangerous. Even if you've done something for years, don't let overconfidence put you in harm's way. Stay humble and stay sharp.

Accidents can change your life in an instant. Never take your health or abilities for granted; protect them, because they can be gone in a second.

Your dream job isn't worth losing to reckless choices. Be mindful of how your actions outside work reflect on your reliability and trustworthiness on the job.

Loss means change, not giving up forever. Adjusting how you engage with your passions will allow you to keep them in your life.

Don't let fear rob you of living. Injuries and responsibilities may change how you play—but don't let them stop you from finding joy.

Own your choices. As an adult, no one will clean up your mess for you. Take accountability and make decisions you can stand by.

Enjoy the ride. Life isn't just about work. Don't get so caught up in responsibility that you forget to actually live it.

Final Thoughts

Growing up, we throw ourselves into the things we love without a second thought, because that's what childhood is all about. But as the adulthood sets in, we start believing that fun has to take a back seat to responsibility.

We stop doing what lights us up because we're afraid of getting hurt, afraid of losing what we've worked so hard for. I've been there. I almost let that fear steal a part of who I am. But here's the truth I learned the hard way: Your job, no matter how much you love it, is not your entire identity. You can't let your career be the only aspect of your life that defines you, or you'll wake up one day wondering where you went.

The lesson isn't to give up your hobbies or passions, but to be smart about how you approach them. You don't have to quit what makes you who you are; you just have to slow down, take a step back, and find that middle ground where joy and responsibility can coexist. The people who care about you want to see you happy, not burnt-out or broken.

Life's too short to trade happiness for fear. Don't get so caught up trying to prove yourself at work that you forget to live. Make memories, chase the moments of joy, and remember what it feels like to be fully alive. Because at the end of the day, that's what makes all of the work worth it.

Reflection Prompts

- ✠ What hobbies do you enjoy?

- ✠ Is there anything you did growing up that you stopped doing once you got older? Why?

21

FUELING THE FIREFIGHTER

Fitness is not about lifting weights or doing cardiovascular exercise; instead, fitness is primarily about maintaining a good diet. In short, what you fuel your body with matters most, and this concept is termed *functional fitness*. Stay committed, stay consistent, and you'll see the results. I am by no means a nutritionist or coach, but the tips in this chapter have helped me to lose about 30 pounds since I began the job.

April 2, 2025

When I started as a career firefighter, my waist was 36 inches, and I was 5 ft 9 inches tall, weighing in at my heaviest of 205 pounds. Today I stand at the same height with a 29-inch waist and a weight that fluctuates—because of burning fat and building muscle—between 175 pounds and 185 pounds. I feel the best I have ever felt.

When I started, I also couldn't do a single push-up or pull-up. That was embarrassing! I could pull heavy weight from a deadlift, but anyone can pull heavy weight if they have high enough body mass index or do it regularly. There's a distinction between being strong and being functional.

Maybe you can pull three or four plates deadlifting 315 or 405 pounds 3 feet off the ground, but can you pick up a firefighter of the equivalent (or greater) weight, then carry them up a flight of stairs because the floor just collapsed beneath them, all after already busting your tail trying to put the fire out for 30 minutes?

No one cares how much weight you can pull if you can't even push or pull your own weight off the ground. Luckily for me, pull-ups were not a requirement for our academy or physical agility test, or I wouldn't be on the job. Now I can do almost anything with my body—pull-ups, pushups, chin-ups, functional fitness, even running. (I can't stand running, but I can do it without dropping, which is a big improvement from my academy days.)

In this job, strength training matters, but functional fitness is greater. So if you have a poor diet, you're screwing yourself! I know some individuals who can do 15 or 20 pull-ups in one shot but can't walk more than 20 feet with two kettlebells doing farmer carries without having to stop. You can be the strongest person in the gym, bench-pressing 225 pounds for multiple repetitions, but if you can't do 15 minutes of as many reps as possible (AMRAP), because you don't have the stamina to keep going, then you're setting yourself up for failure!

How to Achieve a Better Lifestyle

Good nutrition is crucial in this line of work. You don't want to be that firefighter who gains 60 pounds within the first

5 years on the job and loses their functional fitness. The tips here have worked for me, and I'm confident they can work for you too, as long as you stay consistent.

The Basics: Amount, Type, and When to Eat

The total amount of food you consume over the course of the day or week is the number 1 factor in determining whether you gain or lose weight. It's simple: If you take in more calories than you burn, you will gain weight. If you consume fewer calories than you burn, you will lose weight. And if you eat roughly the same amount of calories that you expend, your weight will stay the same.

The type of calories you consume will impact your overall health, energy levels, cholesterol, hormone balance, nutritional status, and body composition.[1] In short, there are healthy and unhealthy calorie sources. If your goal is to gain muscle rather than fat or to lose weight while maintaining muscle, the quality of your food choices is critical. For example, 500 calories from a hot fudge sundae will have a very different impact on your body than 500 calories from grilled chicken, brown rice, and a salad with olive oil dressing.

Ideally, career firefighters should be eating five meals per day. This breaks down as three main meals—breakfast, lunch, and dinner—with (at least) two small snacks in between.[2]

---

1. Frank B. Hu, "Dietary Pattern Analysis: A New Direction in Nutritional Epidemiology," *Current Opinion in Lipidology* 13, no. 1 (February 2002): 3–9.
2. Hassan S. Dashti, "Small, Frequent Meals," American Society for Nutrition, November 11, 2025, https://nutrition.org/small-frequent-meals.

Tips

1. *Focus on quality.* A calorie is not just a calorie. Quality matters. For example, 300 calories from ice cream is not the same as 300 calories from lean protein and complex carbohydrates. Quality foods include the following:

    - Proteins: Meat, eggs, Greek yogurt, cottage cheese
    - Unrefined carbohydrates: Vegetables, fruit, sprouted bread, oatmeal, quinoa, grits
    - Healthy fats: Egg yolks, extra-virgin olive oil, coconut oil and coconut milk, nuts and seeds, almond butter, fish oil, pumpkin, cheese (in moderation)

2. *Eat smaller, more frequent meals.* Eating smaller, more frequent meals can lead to increased energy, better body composition (more muscle, less fat), and improved focus.[3] Start with three main meals and add two to three snacks in between.

3. *Include protein in every meal.* Consuming protein, even in small amounts, can boost your metabolism and

3. Kazunori Ohkawara, Marc-Andre Cornier, Wendy M. Kohrt, and Edward L. Melanson, "Effects of Increased Meal Frequency on Fat Oxidation and Perceived Hunger," *Obesity (Silver Spring)*, 21, no. 2 (2013): 336–43, https://pmc.ncbi.nlm.nih.gov/articles/PMC4391809/.

keep you feeling full longer than sugary, high-carbohydrate snacks.

4. *Meal size should reflect activity.* Your meal size should meet the needs of your activity level. For example, breakfast should be larger since you have a full day ahead of you. In contrast, dinner should be smaller, especially if you're winding down for the day. The following is an example daily meal plan:
    - Large breakfast
    - Small snack
    - Moderate lunch
    - Small snack
    - Small dinner
5. *Time your nutrients.* Consume carbohydrates in the morning or early afternoon or after exercise, when your body can handle them better. Healthy fats should be eaten at night to provide sustained energy and reduce the likelihood of storing excess fat. Protein should be consumed throughout the day.[4]
6. *Consistency is key.* One perfect day followed by one horrible day equals no progress. It is crucial to stay consistent with healthy eating habits.

---

4. John M. de Castro, "The Time of Day and the Proportions of Macronutrients Eaten Are Related to Total Daily Food Intake." *British Journal of Nutrition*, 98, no. 5 (2007): 1077–83, https://doi.org/10.1017/S0007114507754296

7. *Plan ahead.* Plan your meals for the week in advance. Don't go grocery shopping on an empty stomach; you'll be more likely to buy unhealthy snacks. Stick to what you need and prepare meals ahead of time when you know your schedule will be tight.

8. *Reward yourself.* Cheat meals are OK in moderation. Ideally, plan them before or after a workout. If you're eating 35 meals a week, aim for at least 33 to be healthy, leaving room for 2 cheat meals.

9. *Track your progress.* Keep track of your goals and your progress. It's easy to let weight creep up without realizing it. Monitor your progress to ensure you're moving in the right direction.

10. *Enjoy the journey.* Enjoy the process of working toward your goals. Surround yourself with positive people, learn to enjoy exercise and eating right, and remember that progress isn't always linear. Be patient, and appreciate the changes you're making for a healthier lifestyle.

Flashback: The Shiny Bald Head

The 6 weeks leading up to the academy became the catalyst for my fitness success there, all thanks to a fellow firefighter and close friend named Ott. We became friends earlier in the year, though we met in a unique way. . . .

July 2022, 0800 hours

Rush hour on a sunny Tuesday morning, the busiest time in town—school buses, commuters, you name it—and traffic was chaos. We'd already wasted at least 5 minutes attempting to make it through lights. At the time, I was a heavy equipment operator for the town, riding shotgun with a coworker. As we slowly made our way along the backed-up roads, I spotted something out of the ordinary: a shiny bald head, belonging to a man standing by the side of the road, during the peak of morning traffic, dressing a fire hydrant to fill his backyard swimming pool, 200 feet away.

"Am I seeing this right, or am I just imagining it?" I asked my colleague. I recognized him as one of the firefighters from the neighboring town, but I'd never actually met him. A passerby must have called it in, because when we got back to the department of public works yard—the same place the water department runs out of—our fire prevention captain came zooming in looking for the water superintendent. When he couldn't find him, we were the first ones in his line of sight, so he walked straight over and asked if we'd seen anyone dressing a hydrant. That's when he told us the guy had been using it to fill his pool, and the police were already on their way to bag him. Not wanting him blindsided, I reached out through a mutual friend who I knew worked with him often. At the time, I didn't even know his name—just that shiny bald head—but my buddy did. That's when I first learned his name was Ott, and my message was received. Later on, I was given Ott's number and told he wanted to speak with me.

A few days later, I called. Turns out, he wanted to thank me. We got to chatting, and soon enough, I was telling him about my plans to go full time, since I was still a volunteer firefighter at the time. (A month later, I got the job and was accepted into the full-time fire academy.)

Ott mentioned being a personal trainer at a gym, and invited me to come by. I decided to give it a shot. His training focused on firefighter health, fitness, and functional skills, which was exactly what I needed to get ready for the academy. Those sessions turned my workouts around, and pretty soon, I became a beast thanks to him. Ott became a lifelong friend and mentor.

Key Takeaways

Sometimes sacrificing a little bit of strength for more physical agility is what you're going to have to do to become a better firefighter. Yes, it sucks to lose strength, but it's only temporary. You could work out every single day, but if you aren't following a good nutritional regimen, you won't see results. By contrast, someone who works out 4 days a week but is solid with their diet will see more results in just a short time. What might take you 7 months otherwise could be accomplished in only 30 days with a focus on functional fitness.

If your fitness plateaus, usually that will be due to your nutrition. As another of my trainers, Firefighter/Paramedic Scott Barger, taught me:

> *A diet shouldn't be viewed as a punishment;. It should be viewed as a lifestyle! If you aren't having fun with it, you'll drive yourself into the ground. The mind is your biggest obstacle you're going to face head-on. What matters is how you feel so if you feel great then you look great. Don't fixate on the numbers and trust the process.*

"What the heck were you thinking, tagging and dressing a fire hydrant in broad daylight, during rush hour, to fill your pool?"

He giggled, replying, "I went to court, paid the fee, and that was that."

My response was to thank him for tagging that hydrant: "Everything happens for a reason. We wouldn't have met if it weren't for your brilliant plan to use a hydrant in a town you don't work in."

A story of a lifetime, a new lifestyle, and a new friend.

Final Thoughts

Beginning my fitness journey, Ott was my right-hand man when it came to functional fitness. He was the first one to meet me where I was, push me past my comfort zone, and hold me accountable when excuses tried to creep in. Without him, I wouldn't be in the position I am today—30 pounds down, inches off my waist, and with a cardiovascular endurance level I never thought I'd see again. He lit the spark.

But a spark only matters if someone fans it into a fire—and that's where Scott Barger came in. Scott is a firefighter and paramedic in California, a personal friend of mine, and someone I'll always be thankful for. The owner of American Made Coaching, he showed me what it means to set the standard, not just meet it. Scott didn't just hand me workouts—he taught me to love the grind, to push through on the days when motivation was nowhere to be found, and to make health and fitness a complete lifestyle.

Ott gave me the ignition. Scott gave me the drive to keep the engine roaring. Together, they helped me become the firefighter I'd trust to come save my own family. A year ago, I couldn't say that. Today, I can say it with confidence: I'm not just in the fight—but I'm built for it.

If you find yourself stuck or spinning your wheels, don't see it as weakness—see it as an opportunity. The best athletes, firefighters, and leaders all have coaches. Sometimes the smartest, strongest move you can make is to invest in a trainer who can push you, guide you, and pull out the best version of yourself. You don't have to fight the battle alone—success leaves clues, and often, it starts with asking for help.

Reflection Prompts

- What do you do best with your nutrition, and where is there room for improvement?

- How do you feel your diet affects your ability to perform your job?

22

A PROBIE'S GUIDE TO SHIFT MEALS

Meals bond you with your shift by bringing the crew together. The fireground, your work ethic, and your attitude contribute to that as well, but a big crew bonder is the kitchen—where all the magic happens. What may not seem like a huge deal is much bigger than you realize. When you put forth effort into cooking for the crew, it means a lot, even if you aren't a five-star chef. There will be some nights you don't eat and others when you order take out because it's just that kind of shift. Nevertheless, stepping up to cook is a major contribution in this line of work. After all, you spend more time with your shiftmates than you do with your own families, so make the best of it.

After the worst call of your career, some might self-isolate; however, there's nothing like returning to a fresh home-cooked meal made from scratch, with real love, by one of your brother or sister firefighters. The kitchen table is where the chaos comes to an end, and those short moments eating together, enjoying a delicious meal brings so much peace.

Learn to cook, even if it's only one meal. Do your part; people notice. Don't be the probie who burns the food. If you can't be trusted to cook, you can't be trusted to work.

And there's no one as grumpy as a firefighter who has to eat something half-alive because you didn't know chicken had to be cooked to 165°F. I've heard horror stories from friends about uncooked ham, chicken, you name it!

Like everything in this career, taking pride in what you do starts with the small things. Cook enough so everyone can have seconds—and don't forget dessert! Even one solid meal that you know how to make can bring your shiftmates joy, whether after a good or a bad shift. If one suffers, we all suffer together. Lead by example, put forth the effort, and watch the ripples as the positive effect spreads and comes back to reward you. There is no higher compliment than being the chef who did a good job on tonight's meal.

There's nothing worse than a probie saying they'll cook and making something that's not edible. If you do that, then no one will ever want to attempt to eat what you make again—and the jokes will go on for your entire career: "Don't let O'Donnell cook." By contrast, once they see you can dish out a few solid meals, if you mention anything cooking, they'll jump right into an enthusiastic "Heck, yeah!"

To save yourself the headaches, stress, and teasing that follow in the wake of terrible cooking, this chapter shares a few all-around recipes that will get you through your first year. These recipes are intended to feed six on a shift. For a bigger shift, double the ingredients. Hopefully, these meals will earn you a special place in the hearts and bellies of your shiftmates. I'm no expert, but these are my best dishes, which have all had positive reviews. You're welcome! But don't expect these to be the only ones you ever make; these will get you by, but learn to cook at least five more really good meals of your own.

The Salty Rookie's Breaded Bomb Chicken

Prep time: 20 minutes. Cook time: 40 minutes. Total: 1 hour.
Yield: 5–6 servings.

Ingredients

3 pounds chicken breast (6 breasts, about ½ pound each)

2 cups yellow Spanish rice (or your preferred rice)

3 bunches asparagus (about 3 pounds)

½ cup bread crumbs

½ cup grated Parmesan cheese

½ lemon worth of juice

Olive oil

1–2 teaspoons of the following spices, to taste: Himalayan salt (or sea salt), garlic powder, Italian seasoning, smoked paprika

Directions

1. Preheat oven to 350°F. Prepare the chicken by stabbing each breast with a fork to tenderize. Lightly coat a sheet pan in olive oil so that the chicken won't stick. Coat each breast in olive oil, then season both sides with half of the spice mix (reserving the other half for the asparagus). Finish with bread crumbs and be sure the sure the oil soaks into the seasoning (so it's no longer a dry rub; otherwise your chicken will come out like cardboard).
2. Trim the base of the asparagus spears by removing at least the bottom 1 inch (so you won't bite into the tough parts). Lightly coat a sheet pan in olive oil so that your asparagus won't stick. Drizzle with olive oil, then season with the remaining spice mix. Finish with Parmesan and lemon juice.
3. Put the chicken and asparagus in the oven for 40 minutes. While they're cooking, prepare rice per the package instructions, ideally timing the rice to be done at the same time as the chicken and asparagus. Serve with butter once everything is ready.

O'Donnell's One-Pot Pasta

Prep time: 30 minutes. Cook time: 30 minutes.
Total: 1 hour. Yield: 5–6 servings.

Ingredients

2 pounds rigatoni pasta

1 large jar (~24 ounces) vodka sauce

4 ounces heavy cream

8 ounces shredded mozzarella cheese

½ cup fresh Parmesan cheese

3 tablespoons fresh basil (chopped), plus additional for garnish (optional)

2 packages sweet Italian sausage (at least 12 links)

12 ounces 90% lean ground beef

Directions

1. Bring water to a boil in a large pot (make sure to choose a big pot). Add pasta once a rapid boil is reached, and cook until al dente (follow package instructions for cook time).
2. In another large pot, sauté the beef and Italian sausage over medium heat for about 8 minutes, until cooked through, chopping the sausage into small bits as it cooks. Drain fat and oil once cooked. Ideally time it so that the pasta and meat are done around the same time.
3. In a separate pan, add vodka sauce over low heat. Stir in heavy cream until combined, then add diced fresh basil, Parmesan cheese, and mozzarella cheese. Simmer for 15 minutes, stirring every 3–5 minutes.
4. Once cooked, drain the pasta and dump back into original pot. Add in the cooked beef and sausage, then slowly pour the vodka sauce over everything. Mix until well combined, then serve (with optional basil to garnish).

The Salty Rookie's Fall-off-the-Bone Firehouse Ribs

Prep time: 35 minutes. Cook time: 2 hours 25 mins. Total: 3 hours.
Yield: 5–6 servings.

Ingredients

4 pounds baby back pork ribs

2 tablespoons olive oil

For the spice mix:

2 teaspoons garlic powder

1 teaspoon onion powder

2 teaspoons paprika

3 teaspoons salt

1 teaspoon cracked black pepper

½ teaspoon cumin

1 teaspoon chili powder or cayenne pepper (optional if you want a kick)

For the sauce:

2 cups barbecue sauce

2 tablespoons olive oil

3 tablespoons minced garlic

1 tablespoon Worcestershire sauce

½–1 tablespoon cayenne pepper (optional for heat)

1 teaspoon salt

Directions

1. Preheat the oven to 350°F.
2. Peel off the tough membrane that covers the underside/bony side of the ribs. Place on a baking sheet or tray lined with foil (or parchment paper).
3. Coat the ribs with cooking oil, rub the spice mix all over, then finish with a light drizzle of oil on top.
4. During the last 5 minutes of cooking time, mix together sauce ingredients.
5. Remove ribs from the oven, remove foil and spread the tops of the ribs with the barbecue sauce mixture.
6. Increase oven temperature to 460°F. Return ribs to the oven, uncovered, and bake for a further 10 mins. Change oven settings to broil (or grill) on medium-high heat to lightly char and caramelize the edges (about 3 mins).

Rest for 10 minutes to allow the juices to recirculate back into the meat before slicing to serve. Enjoy!

O'Donnell's Breakfast Tortillas

Prep time: 15 minutes. Cook time: 23 minutes. Total: 35 minutes.
Yield: 12 tortillas (5–6 servings).

Ingredients

1 dozen 8-inch tortillas

16 ounces sausage

12–16 ounces bacon

8 ounces shredded cheese

1 dozen eggs (1 for each tortilla)

½ cup chives, chopped or diced

16 ounces half-and-half or cottage cheese (optional for eggs)

Salt, pepper, and your preferred egg seasonings, to taste

Directions

1. Preheat oven to 350°F.
2. Put tortillas in a muffin sheet. (Stuff each to form a large tortilla chip.)
3. Precook bacon and sausage (until almost completely cooked).
4. Dice your bacon and sausage into little pieces, then set aside.
5. Whisk 1 or 2 eggs in bowl. Add seasoning of choice and half-and-half.
6. Pour mix into tortilla (2 eggs for each tortilla).
7. Add as much bacon and sausage as fits into each tortilla along with eggs.
8. Top off mixture with shredded cheese.
9. Garnish top with chives.
10. Put into oven and bake 23 minutes (or until egg puffs up or over; you'll be able to see it).
11. When done, take tortillas out of oven. Let sit 5 minutes before serving.

Enjoy! Two to three tortillas per person should be more than enough to satisfy hunger. You're going to want a knife to cut into these. They'll be hot and good.

Homemade Chocolate Chip Cookies

Prep time: 15 minutes. Cook time: 11–12 minutes. Total: 30 minutes. Yield: 36–40 cookies.

Ingredients

- 2¼ cups flour (unbleached, all-purpose, or preferred flour)
- 1 teaspoon baking soda
- 1 cup (2 sticks) salted butter, plus additional to grease pan
- ¾ cup brown sugar
- ¾ cup white sugar
- 2 eggs
- 1 tablespoon vanilla extract
- 12 ounces of chocolate chips of your choice (semi-sweet preferred)

Directions

1. Preheat oven to 350°F
2. Mix the flour and baking soda in a bowl and set aside
3. In another bowl, cream 2 sticks of butter.
4. Beat in 3/4 cup brown sugar, and 3/4 cup white sugar, until combined.
5. Add in 1 tbsp vanilla extract, then add in eggs, one at a time. Once these are completely combined, add in the flour and baking soda mixture. Beat until just combined (consistency should be like playdough; don't overmix or it'll get soupy). Finally fold in chocolate chips until evenly distributed.
6. Grease the bottom of a sheet pan with butter (just enough so cookies don't stick to pan).
7. Form dough into tablespoon-sized balls. Place dough balls 1–2 inches apart on greased sheet pan.
8. Once the pan is full, put cookies into the oven and bake for 11 minutes. Note: Cook only one sheet at a time. The cookies will not cook evenly if you do more than one.
9. Take out of the oven. Let cookies sit on sheet pan for 5–10 minutes before removing from the tray with a spatula. (The cookies are still baking on the pan when you take them out. This step is important if you want a soft chewy bite without undercooking. If you rush the process, the cookies will not be cooked long enough.)

After cookies have cooled, enjoy!

What's your favorite recipe? What does everyone ask you to bring to cookouts and parties? Use the space below to jot down your favorite recipe, whether from your family or your own creation.

Recipe Name:______________________________

Prep time:________. Cook time:________. Total:________.
Yield: servings:________.

Ingredients

Directions

Reflection Prompts

- ✠ What's your preferred comfort food after a rough day?

- ✠ What's your favorite meal to cook for your department, and why?

23

PROBIE TO PROUD

During the probationary year, our department conducts quarterly reviews to make sure probies meet the expectations of the job. To go beyond my own perspective, I want you to see exactly what my captain wrote about me at this stage, along with my reflections. It's one thing to tell my story; it's another to give you a glimpse of how I was seen by those who evaluated my growth. I believe in transparency, to present a fuller, honest picture of what this journey looks like from both sides of the helmet.

First-Quarter Evaluation

April 16, 2024

> *Probationary Firefighter O'Donnell has come to us as a lateral transfer from the highway department. Nichole has some experience in the fire service and also very limited EMS experience as she has obtained a BLS [basic life support] license but never had worked with*

a transporting service. Nichole has already attended the [Massachusetts Firefighting Academy] recruit program in which she did very well. I had received reports from training staff and she has picked up the trade and understands all the basic requirements of a day-to-day firefighter. Nichole initially came to B group in early February when I was not the shift officer. The month I was not present, I received the following report: "Nichole has started to work through her probationary binder and is starting to be accustomed to the shift work and life. The basic firefighter skills appear to be in check as expected fresh out of the academy but requires a great deal of work on the EMS side as policies, protocols, skills, assessments, and paperwork. The attitude has been great and has that 'can do' attitude."

Since the beginning of March, I have come to see a lot of the same traits. Nichole possesses a very positive attitude and is always trying to keep herself moving in the right direction. Communication skills have been good and ask questions appropriately. Basic fire skills appear to be most proficient and only lacks experience which will be gained over time. The EMS side will be a work in progress and will be for some time. This is mostly due to the lack of experience, not the lack of trying. Nichole does ask questions and appears to learn but sometimes requires a refresher. Again there is a lot of material to take in for the day-to-day work and any deficiencies are addressed and corrected [promptly].

Nichole has fit into B group very well and understands the role as a probationary firefighter. Always staying busy and always trying to better herself with either fire-based skills or EMS skills. Nichole always is willing to clean and keep everything in the station tip top which is a great help to all shiftmates. At times Nichole had a difficult time just standing back and allowing members to take their role and understanding we all have one job to do and not allow cross over. This is just part of the go go go attitude that Nichole expels. That being said it is an excellent trait to have, and it just needs to be controlled.

The plan moving forward will be to continue fire-based training and with an emphasis on riding position and related topics. Also to assist as much as possible with EMS skills. Nichole has also just signed up for medic school and she will start in the fall. The plan is to make sure she is proficient with all BLS skills before starting school again. The overall evaluation for Nichole is good and it has been a pleasure to work with her to date. I look forward to watching her grow with the department and continue to achieve her goals.

April 12, 2024

A lot was thrown at me during those first few months of probation. Heading into quarter 2, I'm patiently waiting for the next review, and this time, I know there will be things we'll need to address. That's part of the process. Probation is about making mistakes, learning from them, and coming out stronger on the other side. It's about working through the rough spots, pushing yourself, and building a solid foundation so those mistakes don't happen again.

I had a picture in my head of what this journey would look like, but reality has been different in ways I didn't expect. The job itself isn't impossible, but it demands your full focus, effort, and commitment every single day. In turn, it gives back what you put into it.

Now I look forward to those check-ins with my captain, whether it's to talk about areas I need to work on, constructive feedback, or signs of growth. Because one day, I'll be on the other side, helping the next probie to navigate their own challenges, using what I've learned to guide them through.

August 1, 2024

This year has flown by. Eight months ago, I walked out of the academy triumphantly, and the time since has been a blur. Probation is almost over, but I still feel that I'm not ready to move on. The weight of new responsibilities is starting to hit harder, especially now that I have a junior member under me.

Ten months in, I've reached that point where my shiftmates are finally comfortable enough with me to bust my chops. It's a small but important sign that I've earned my place. Going from probie to part of the crew takes time, like everything else in this job.

With paramedic school just weeks away, my focus is shifting toward that next challenge. I'd be lying if I said I wasn't nervous, because I am terrified of failing and unsure what to expect. But I know hard work, persistence, and dedication will carry me through, as they have so far.

I've come a long way as a basic emergency medical technician (EMT). I'm no expert—not even close—but I'm better than when I started. I remember freezing up trying to work the stretcher, stumbling through patient contact, needing help on every call. Now, I can do assessments, pick up on subtle signs, assist medics, and now tech my own calls. The growth occurred, even when I didn't always see it in the moment.

The biggest takeaway here? Skills can be taught. What matters most is attitude. A positive, eager, and teachable mindset will take you further than talent alone ever could. If you're new to the job, remember that your willingness to learn earns you respect and shapes the firefighter you'll become. Keep pushing, and trust the process. After all, everyone started somewhere.

Second-Quarter Evaluation

August 20, 2024

This is the second quarterly review for probationary firefighter O'Donnell as she was at the fire academy upon being hired. Nichole continues to maintain a great positive attitude towards not only shift work but the service we provide to the town. Following up from our first review we have continued to work towards strengthening Nichole's BLS skills. This has been a continued process of review and correction. The lack of EMS experience has made the process somewhat difficult, but we continue to push forward to have Nichole gather the skills and experience to become a better BLS provider.

Nichole has such a strong attitude to being the best she can be; it leads to being extremely nervous to make mistakes. We have reviewed that mistakes will happen and that is part of the learning process and no one is perfect.

The medical side of the job has started to come around for Nichole since our last review. Leading Nichole into the lead role of an EMS call has helped her start to understand the flow of the calls and how each one shall be run. Nichole takes direction well and applies it to the calls. It has been noticed on calls that if conditions change rapidly there is some difficulty with adapting and Nichole becomes overwhelmed. Bedside manners for the most part has been OK and just keeping it professional and with reduced intensity has been discussed.

Nichole has continued to take part and do very well with in house training. The one thing we are also trying to focus on is not getting ahead of herself and focus on BLS. This will be difficult as Nichole will be starting paramedic school next month. We need to focus on maintaining the very well-structured fundamentals of the BLS program so that Nichole has the skills to succeed in her paramedic class. There is some concern that Nichole may not currently have the tools and experience to have a favorable outcome in paramedic class, but I think her strength of determination of success will prevail. This will be monitored very closely.

As stated in previous evaluation, Nichole is progressing well. Always staying busy with station duties and or educating herself. Moving forward, our plan will be to continue to focus on EMS skills as the basic firefighter skills are in line for current position and timeline. I think that monthly reviews should be done moving forward to confirm that paramedic school is on track and there are no other areas that need to be addressed.

August 28, 2024

That went much better than expected. I always put so much pressure on myself, convinced I've messed something up or fallen short. But this review? It wasn't anything like I'd feared. The mistakes I thought were career-ending weren't even mentioned—they're just part of the learning process now. I've adapted, and now it's about continuing to progress without letting frustration take over when mistakes happen—letting go of the pressure to be perfect.

This job teaches you something new every day. Perfection would mean you've got nothing left to learn, and that would make you unteachable. I'd rather be the one who stumbles but grows than the one who coasts without growth. The goal isn't perfection; it's being coachable, being willing to learn.

Early on, I tried to be the perfect probie. Consequently, I was so focused on not screwing up that I ran myself into the ground. My stress went through the roof, my heart rate spiked on every call, and I couldn't concentrate. I didn't have balance and had forgot that meeting expectations isn't about being flawless, but about being dependable, learning, and improving. Overcoming that perfectionist mindset didn't happen overnight. It took time, and it is still a work in progress. You can't avoid mistakes, and you can't skip the process. Instead, learn to expect bumps along the way, and stop letting them shake you so much. Next time a tone drops, take a breath and say, "Challenge accepted." You can't control every outcome, but you can control how you respond.

What sets the experienced apart isn't talent or luck; it's time and what they did with it. One day, I'll be the one others look to for guidance. And when that day comes, I'll remember that making mistakes isn't the problem; not learning from them is.

Bittersweet and Humbling

October 4, 2024

Six weeks into paramedic school, it's already been a journey of growth. My doubts have given way to quiet confidence. I'm not just learning the material; I'm learning who I am. College opened a part of me I hadn't tapped into, showing me what's possible when I stop doubting and put in the work.

On the job, I feel it too. I'm not the uncertain probie I once was. The confidence I have now comes from embracing mistakes and growing through them. Small milestones—like earning my nickname last week (see chapter 25)—remind me how far I've come. This year taught me more than I ever imagined, and though it's bittersweet to see it winding down, I'm proud of who I have become.

Containing My Excitement

October 31, 2024

I can't even put into words how surreal this feels. Fire Engineering Books believed in me right from the start just like Captain Arventos did. Today, I got my contract, and I'm stoked. I started writing this book back in May 2024, pouring my heart onto the page, sharing the lessons I've learned, the struggles, the victories. And 6 months later, that idea I wasn't even sure I'd follow through on has turned into a manuscript and become a reality, in one heck of an amazing career.

When you give something your all—when you really believe in yourself and stay true to what drives you—incredible things can happen. This is just the first of many goals I plan to realize. I know without a doubt that I'm capable of so much more.

Paramedic School Progress

November 6, 2024

I'm 6 weeks out from finishing my first semester of paramedic school, and I'm proud of how far I've come. I won't pretend it's easy, but I have learned that you really do get out what you put in. That mindset helped me score 97% on my first examination last week, and while I know not every test or grade will go that way, it showed me what I'm capable of when I focus and trust myself. Anything is possible when you put your mind to it.

I'm also realizing how important it is to find balance—to manage my time, protect my energy, and stay grounded so I don't burn out. The fear that once held me back? That's fuel now. Each step forward—every challenge I take on—is shaping me into the firefighter, paramedic, and person I'm working so hard to become.

Final Probationary Firefighter Evaluation

November 18, 2024

> *This is the final review for Probationary Firefighter O'Donnell, as she was at the fire academy upon being hired. Nichole continues to maintain a very good work ethic at the station and towards calls and training. Training continues to be on the forefront of our two days each week. Nichole continues to use her time wisely, keeping up with medic school studies. At this point, I believe Nichole has fulfilled the basic requirements to pass the probationary period. Nichole needs to stay focused on paramedic school and gain as much experience as she can from every call she responds to. Stay positive, stay determined, stay confident, and you will succeed with everything you do in life and within the fire department. It has been a pleasure assisting Nichole with the probationary year, and we will expect great things from her in the future.*

Key Takeaways

Growth doesn't come from being perfect. Instead, it comes from staying disciplined, teachable, and committed to improving every day.

Your credentials and experience on paper go only so far. Your attitude, work ethic, and how you treat others are what truly define you.

Stepping off probation isn't the end of the hard work. It's the start of proving why you belong.

This career is a privilege. It takes time to earn, but one poor decision can take it away in a heartbeat.

Your impact on others is shaped by how you show up: positive, supportive, and driven—or the opposite. The choice is yours.

Final Thoughts

As my probie year closes, I'm filled with gratitude. What once felt like constant fear, the pressure of evaluations, and the weight of proving myself has been replaced with quiet pride. This year wasn't about being perfect. It was about showing up, staying disciplined, and learning with every challenge. The mistakes, the small wins, the hard lessons—all have been stepping stones in becoming the firefighter I want to be. After years of struggle, setbacks, and growth, I made it.

Securing my dream job is one of my greatest achievements, and now I'm stepping into the next chapter, paramedicine. If you're reading this as a probie, take this advice to heart: You can come in with all the credentials in the world, but without discipline, work ethic, and the ability to work well with others, those pieces of paper don't mean much.

The most valuable trait you can bring to this job is being teachable. The calls, the hours—they'll always be there. But if you don't learn how to separate yourself from the job when needed, burnout can happen before your career even begins. Stepping off probation doesn't mean it's time to relax or stop caring; it means it's your turn to show why they made the right choice when they hired you. This lesson applies in any career.

You have the choice to be an asset or a liability, to do the right thing even when no one's watching, and to make this the best career of your life or let it be a decision you regret. Be the firefighter who lifts others up, who checks on your crew, who leads by example. Never forget what it felt like to be in these shoes, figuring it all out. You won't know everything, and that's OK. You have a lifetime to keep learning.

This job is a privilege. One bad decision can take it all away. It takes a long time to earn the opportunity, but only a moment to lose it. Remember that. The way you carry yourself, the attitude you bring—that's what people will remember.

So here's your moment. Step up. Set the standard. Be proud. Give a damn. And never forget how lucky you are to do this job. The ball's in your court now—it's your time to shine.

Reflection Prompts

- What was your experience as a probie like?

- What kind of firefighter (or professional) do you want others to see when they look at you? How can you work toward that vision?

- How will you hold yourself accountable to the standards you want to set on and off the job?

24

THE NEW PROBIE

July 29, 2024

Today marks the official end of my reign as the junior firefighter on the job. We've got a new probationary member starting, and I'm feeling a whole truckload of emotions, mostly excitement, but there's definitely a hint of *Oh man, now I have to be responsible.*

It's a weird mix of relief and panic. On one hand, I finally get to be the one showing someone else the ropes, steering them in the right direction, which is honestly pretty cool. On the other hand, now I'm expected to know better; no more "O'Donnell moments," no more getting away with rookie mistakes.

The day kicks off in high gear, as usual. The new probie's just trying to find his footing. I introduce myself with a firm handshake and try to give off my best "I'm not really intimidating, I swear" vibe. Right away, I like him. He matches my energy and seems eager. Plus, he showed up with doughnuts for the crew. Smart man, he knows how to win friends fast, and I can tell he's got a bright future ahead of him.

July 30, 2024

The new probie's crushing it. He brought in fresh bagels with homemade cream cheese—my favorite. Nothing lasts long around here, especially bagels, so I knew he had it figured out. I'm on a 34-hour shift today, picking up overtime to achieve maximal staffing, and I'm genuinely excited to spend time showing him the ropes.

As we're chatting, I find out he's not a complete rookie. Before this, he had worked for a department about an hour away, and he graduated from the full-time academy a few years back. That experience explains why he seems squared away already. He's not walking in blind like I did. It's a different vibe training someone who's already been through it somewhere else.

I ask, "Hey, did you guys ever specialize in rapid intervention teams or the 2:1 setup?"

He shakes his head, "Not really."

That's my cue: I mention how much we stress training on rapid intervention and search and rescue. I suggest we set up a ladder and set up the 2:1 system. We head outside, and I break it down—the assignment, the why behind it, the method. Then I tell him, "All right, your turn to be the teacher. Show me how it's done."

He nails it on the first try—like he's done it a hundred times. I'm impressed and half convinced he was bluffing when he said he hadn't done this before. We finish the session by diving into our department's gold mine of archived training files. I give him the rundown of how we operate, what makes us different, and what we expect here. We spend a solid 4 hours at it, and before I know it, his shift is done. He thanks me, and I'm genuinely looking forward to working with him more. Today was a good day. And just like that, my time as the junior probie is officially over.

It feels weird. Did I want this day to come so soon? Maybe not. There's comfort in being the probie, because you get room to mess up and learn. Now? It's my turn to step up, to be an example, to act like the mature adult they expect me to be.

January 15, 2025

Six months later, the new probie has been pinned at our annual ceremony. Calling him "new" doesn't feel right anymore. He's one of us and one of the smartest firefighter/paramedics I've had the chance to work alongside. I didn't need to mentor him much because he was self-directed, sharp, and a hard worker. I still get to call him "kid" around the firehouse, though, since even though he's older than me, age doesn't matter when it comes to earning respect. And he's earned mine, as well as the department's.

He has been assigned to another shift, so we don't get work together regularly. But on the occasions when we do, it's enjoyable. It's refreshing to see someone come in with experience, but still humble, hungry to learn, and willing to put in the work. Watching his progress reminded me not only of how far I've come but also of how much I still have to learn.

There's always room to grow, no matter where you are in your career. And this is drives home a point that matters:

Don't confuse time on the job with real experience. There's a huge difference between someone with 20 years in a slow firehouse who's spent most of it parked in a recliner, and someone with 5 years in a busy house who's been through the thick of it. Time alone doesn't mean you put in the work or gained experience. If you ever have the chance to join a busier department—even if it means starting over as a probie again—take it. Go where the experience is. Getting out of your comfort zone is how you grow.

Junior firefighter isn't just a spot in the pecking order. It is a chapter of learning, messing up, and being given grace while figuring out the job. For me, that chapter is over now that the new probie is here; it's my turn to be the example, to step up, to carry myself like the firefighter I looked up to when I started. No more hiding behind being the probie; no more excuses. The comfort of being the new kid is gone,

Key Takeaways

Being the junior firefighter is temporary. What matters is how you grow when it's your turn to step up.

Experience is measured by what you've done and learned, not just by time served.

True leadership starts the moment you stop hiding behind the title of junior firefighter and start taking responsibility for those coming up behind you.

A good firefighter doesn't stop learning, regardless whether a probie or the one teaching.

Respect is earned through work ethic, humility, and how you show up for your team, not by seniority alone.

and that's OK. Because now, I can show up every day having earned my place, while doing my part to help the next probie earn theirs.

Final Thoughts

More than the time you have on the job, the character you hold within will drive your growth in the fire service. Anyone can step up and lead, but if you don't have the right mentality, you're going to find yourself in a pinch.

Even though he had experience, the junior firefighter who I mentored was open minded. He didn't object to reviewing a skill that he already knew to some degree. Being willing to be shown anyways allowed us to explore other aspects of the skill and learn from each other. This falls back into the show-don't-tell mindset. Learn from the new guy.

There's comfort in being the probie, with space to make mistakes and learn under someone else's guidance. Eventually, though, that chapter closes—and when it does, it's on you to step up, to be the one others can count on, and to help the next person along. Experience isn't about how long you've worn the badge; it's about what you've done with that time, how much you've learned, and being willing to keep learning. Titles don't make a good firefighter; attitude, effort, and the drive to earn respect do.

Reflection Prompts

- How long was it before you had a junior firefighter under you?

- What did it feel like when you advanced beyond junior firefighter?

25

A YEAR OF FIRE, GRIT, AND GROWTH

September 25, 2024

One year later, I remember well the day I walked into that interview with the selectboard. Nervous, but confident this was exactly where I was meant to be. The intervening time feels like the blink of an eye.

Everyone told me the first year would fly by, and they were right. Seems like yesterday I was a brand-new probie, and now I'm a year into the job. Time has a way of slipping past you when you're caught up in the daily grind, but the reality of how fast it passed shocked me. If a year could go by this fast, what would the next decade feel like? By then, I'll have a dozen new firefighters looking up to me for guidance.

One Day at a Time

This year was a blur. From the highs of being on the nozzle at my first fire to the lows of burning out, the memories began to blend together. Coming up on 4 years in the fire service, I remember clearly my first big mistake—the one that hit me like a ton of bricks thinking I'd never recover from it—but others were fuzzy around the edges. Which is why it's so important to write things down—to not forget the lessons learned.

Advice from an Overeager Probie

Looking back to the day I was hired, I remember exactly how I felt. Even with everything this year has thrown at me, I still feel the same way. During my candidate interview, I asked the chief, "What's the one piece of advice you'd give to me or anyone new to this profession?" His reply: "Don't lose the feeling you have now." I couldn't agree more, and that one piece of advice will stick for my career.

If you love what you do, you'll never work a day in your life. That goes for any job, not just firefighting. You aren't getting any younger, and the days go by and the world keeps spinning, whether you're present or not.

Make every day your best, treat it like your last, and take it one day at a time. I am going to miss being a probie. Still, the clock keeps ticking, and that's why living in the present is so important. If you're always looking forward to the "I can't wait" stuff, you'll miss out on the present; the years fly by, and those "I can't wait" moments turn into "Remember when" instead. Ultimately, everything shifts to "Where did the time go?" So pause and take in the present moment. At some point, you'll find yourself wishing you could get back the time you had, regretting that you took life too seriously.

Make the Best of It

September 27, 2024

I work with a senior member who's getting ready to retire within a few years. I give him a run for his money when it comes to talking—once I get going, I don't stop. Once, chatting with another private, all energy, I had spit out probably 1,000 words in a single sentence. The senior member, overhearing us, yelled at me, "Nichole, breathe! Did you just snort a line?" Cackling with the other private, I watched this guy practically pull out what little hair he's got left. I kidded him, "Give me 24 hours in a room with you, and we'll have your retirement paperwork filled out." The other private, laughing, chimes in, "Try 30 minutes."

One day, you'll be that senior member almost ready to retire, and some upstart probie just like you will come along. You'll just want to enjoy a cup of coffee, but they won't shut the hell up and give you 5 minutes of peace. I usually give my crew their peace, aside from the spills, trips, falls, and probie chronicles, but for this one senior member in particular deep down, I know he secretly loves it—even if he'll never admit it. Life's too short, and sometimes you just gotta hit 'em with a Donald Thornberry laugh.

Keeping Yourself in Check

October 2024

I've realized that 1 year as a probie just doesn't feel like enough. Honestly, I don't feel ready to come off probation yet, not because I'm not capable, but because a year doesn't give you nearly enough experience in this job. Every call is different, and each one teaches me something new; still, there's a lot I haven't faced. That feeling stays with me—that I've come a long way, that I've earned some of my confidence, but that there's still a gap between where I am and where I need to be.

A year is just a drop in the bucket when you think about all the situations that can happen on the job. I've had my share of tough lessons, which have pushed me closer to being ready, but I can't shake the feeling that I haven't yet hit all the marks. Maybe that's my past creeping in, feeling like nothing is ever good enough, or maybe it's my mind getting in the way, second-guessing myself.

The truth is, I know I'm not the captain's "best" on the crew. I won't hesitate to admit that the more seasoned crew members are better than me. They have more years on the job, more calls behind them, and have faced situations I haven't experienced. I'm at the bottom, and I own that. It's not for a lack of effort; I show up every day ready to give 100%. I know where I stand, and I'm OK with that; it's part of the process while I'm earning my stripes. Inexperience takes time to overcome.

Even though I'm not the best, that doesn't mean my crew doesn't trust me. There's no questioning my willingness to step up and handle my job; if I'm given an order, I'll execute it without hesitation. Strength isn't an issue either; without a doubt, I could pull any one of my crew out of a burning building if I needed to.

I'm self-aware enough to speak up when something is beyond my abilities. I'm not reckless. I know when to push myself, but I also know when to ask for help without letting my pride get the best of me. When it comes to more complicated tasks, I'm not the first one they turn to. And again, that's not because I'm lacking intelligence, it's purely down to experience.

You can't expect someone who's new to the job to solve every problem that comes their way, especially not the ones that require years of knowledge and on-the-job learning. That'd be like asking a civilian to pump the truck when they don't know what a nozzle is. It's not realistic, and it's not fair to expect someone in my position to have all the answers just yet. I'll get there with time and more experience, so that one day I'll be the one people look to in those moments. I'm still learning, and I'm OK with that.

Know where you stand. There's no shame in recognizing your weaknesses and acknowledging that it takes time to get where you want to be. Even so, it weighs on me more than I'd like to admit; part of me wants to be the one they turn to, the one they trust to figure it all out. I have to remind myself that this is a journey, and since each call is part of the journey, I'm getting closer. I'm trying to be patient, even though that's not always easy.

Earning Your Place

April 6, 2025

The year that flew by has become one I'll always cherish. When I started at my firehouse, I was the fresh-faced probie who no one knew or trusted. Under constant scrutiny, I was not sure I'd ever fit in. Over time, through trial and error, I transformed from an outsider into someone who had earned respect the right way. With that shift, some of the members who had barely spoken to me started to open up. At first, this occurred in short exchanges of a few words here and there, but by the end of my probie year, there was a genuine connection.

That shift didn't happen overnight; it takes time to build trust, but the slow process is worth it. With 2 years on the job later this year, I now have formed what I call my *fire family*. Which makes me question why some people pursue inappropriate relationships. We're all a big family, how could someone ever want to screw that up? I'll never understand it.

Being the first woman hired on to this department in almost 15 years, I see why they constrained themselves around me. Still, that didn't deter me.

This was a year of growth, evolution, and learning. It was filled with challenges, steep learning curves, and moments in which I had to prove myself. Not because I was a woman, but because I was the probie. No one earns a place in the department without putting forth the effort.

So I kept pushing forward, head down and nose away from other people's business, and I slowly earned the respect of the senior members of my department. I worked hard to get where I am, and I take enormous pride in that. One of the proudest moments of my probie year was finally earning the nickname "N-ROD"; more than a name, it's a sign that I belong. It's also a badge of honor that represents the blood, sweat, and hard work it took to prove myself, not only as a firefighter but as part of this fire family.

For the Haters and Doubters

What drove me this year wasn't just the lessons I learned or the challenges I faced, it was the people who doubted me. The ones who whispered behind closed doors, who said I wouldn't make it, who bet against me like my future was a game. I heard their chatter, even when I wasn't supposed to:

"Good luck. I've been dispatching here for years, and they still won't hire me. Unless you've got your medic and 10 years on the job, you'll never get into that department."

"You better be better than everyone else, because you're a girl. It's going to be a hell of a lot harder for you to get off probation than it was for the others."

"There are bets that you won't make it through your probationary year."

"I lost money because of you—but congrats, I guess."

Those words could have broken me. But instead, they lit a fire under me. Through every shift, every drill, every sleepless night, I carried those words with me—not as a burden, but as fuel. I didn't want merely to prove them wrong; I wanted to prove to myself that I was stronger than their doubts, stronger than the old version of me they tried to pin me down to. I didn't just survive my probie year—I crushed it. No shortcuts, no handouts—just hard work, grit, and determination. Today, I stand tall, with my head held high and with my name stitched across my gear no longer reading Probationary Firefighter O'Donnell, but Private O'Donnell.

To those who doubted me, my success is the best response. This year has been about perseverance and turning all the motivations into positive energy. As at the academy, it's all what you put in: Apply yourself fully, stay positive, and there's no reason you can't succeed. My probie year was one for the books.

Key Takeaways

Your first year flies by, so make it count. Time passes faster than you expect when you're focused on learning and growing. Don't miss the small moments that shape who you become.

Mistakes are part of the process. It's better to fail, learn, and grow as a probie than to fake your way through and pay for it later.

Respect isn't given; it's earned. No matter who you are or where you start, trust is built through hard work, humility, and consistency.

Even doubt can provide motivation. The words of those who didn't believe in you don't have to break you; they can push you to become stronger.

The job is a craft, not just work. Mastery comes with time, dedication, and a mindset of always striving to be better.

Final Thoughts

Your probie year isn't just about proving yourself to others, but about proving to yourself that you belong, that you're capable, and that you have what it takes to do this job with integrity and heart. Every call, every drill, every long night added up to the firefighter I am today: not perfect, but determined, humble, and always ready to learn.

If you take anything from my journey, let it be this: Don't rush the process. Embrace the grind, stay present, and let the hard days sharpen you. Let the doubters light a fire under you, but don't let them define you. You'll never regret working hard and staying true to yourself, and in the end, that's what earns you your place in this family.

Reflection Prompts

- ✠ What did you learn from your probie year?

- ✠ What's the single best piece of advice you would give to a probie just starting out?

26

SOCIAL MEDIA

January 27, 2025

Fresh off probation for a couple months, we got dispatched for a fire, and I was the pump operator. This was not my proudest moment (and even though technically outside my probationary year, it ties into the themes of this book). I had only one job: to stay at the pump, with the truck.

Once the fire was extinguished, I decided to post a story about the fireground itself on Snapchat. No other people were in the picture, and this wasn't a structure fire. Given the circumstances surrounding the fire, though, I could have been subpoenaed. Not only that, my department would take the backlash for my decision to post this. Meanwhile, I just thought it was an average fire.

Won't Do That Again

January 31, 2025

I got pulled into the chief's office for that one. I knew I screwed up, and the last thing I was going to do was hide from my mistake. Owning your screwups is essential. The chief knew I made a mistake, and so did I. There was no running from it, just embarrassment, a written reprimand, and my acknowledgment that "Yep, I won't do that again."

This led me to deactivate all of my social media accounts for 3 months. I deleted anything remotely related to my department, and when I reactivated my accounts, I limited my posts to fitness, lifestyle, and writing.

Going dark from social media was the most refreshing thing I've ever done. I recommend you do the same at least once in your career. No one knows anything that's going on in your life, and the peace that comes with it is so great!

More Harm Than Good

I understand the urge to tell everyone that you just got hired onto a fire department full time and the temptation to post about exciting moments on the job. That instinct reflects how much this job means to you. But the way of the world is that one wrong post or one careless comment—or even liking the wrong content—can hurt you.

What Not to Do

Don't be the probie who blasts their job all over social media in their first year, save for one exception: when you graduate from the fire academy. That's your personal milestone—be proud of it and share it. Not everyone can accomplish what you've done, and graduating deserves recognition.

Other than that, keep the job off your feed. Posting pictures from calls, selfies in gear, or anything that hints at incident details can violate department policy, compromise patient privacy, and make you look like you care more about likes than the work. Word travels fast in this business, and the wrong post can follow you for the rest of your career.

This isn't just for rookies—don't post photos of yourself in a department shirt while out partying. Even off the clock, you're still representing your department. Similarly, avoid posting pictures in turnout gear captioned, "fighting a structure fire." What you see as pride, leadership may see as poor judgment or a liability.

And while we're at it—don't saddle your whole identity with the job by being that person who buys a hoodie or T-shirt that screams, "I'm a firefighter," with giant Maltese crosses, axes, flames, or slogans like, "First In, Last Out" plastered across it, and then wears it everywhere like you're a local celebrity. Pride is fine—ego isn't. Let your work speak for you instead of your wardrobe.

Don't use your credentials to get out of things, either. You're a public servant, not above the law, and the badge doesn't grant you special privileges outside of the work you do for your community.

Stay away from posting about politics, especially in local town forums. Don't like or comment on hateful or controversial posts, even if you agree with them—it still puts your name and your department's reputation in the middle of it.

Bottom line: If you have to stop and think, "Should I post this?," the answer is no.

Keep Your Opinions to Yourself

For the record, I don't care for political parties, beliefs, or who likes who. I keep to myself and don't want others knowing about my personal business. We're all entitled to our views on politics, but I am strongly against sharing them. I don't trust the government, but I like the peace of knowing no one hates me because I think differently than they do.

The advice I'm about to give you is based on sound reasoning—namely, you don't want a target on your back in your career—so I strongly encourage you to pay attention. I was fortunate enough to learn this lesson while I was young. Then, I could move forward without it derailing an entire career. We have a first amendment right, but you need to be smart. Union contracts can nip you in the tail, so forget about promotions and good luck keeping your job!

We all have our own beliefs, but get away from sharing them with the world. Regardless of rank, it isn't OK to sound off at or about work. You should never have to worry that someone will say the wrong thing at your workplace, just because one of your members has different political views.

Hypothetical Fallout

Say I decide to engage with a post where a resident is clearly in the wrong, but most of the town backs her up. If I post something unnecessary to make myself feel better or correct the record, you know what she's going to do? She's going to stalk my page, find where I live, and figure out where I work.

If your page is public, she'll know everything about you. Social media is no longer the place to air your opinions, especially if you're a probie.

Let's add to this hypothetical situation that you're a white man and made a derogatory comment elsewhere on your page about a topic like Pride Month (an example chosen since a majority of the world is hot on this topic). That resident from the forum now has the ammunition she needs to take you down. And if your profile picture is you standing in front of the fire truck or wearing department gear, guess who she's going after next? Your department. It's no longer just "John Doe was awful to me," but "Firefighter John Doe was awful to me."

Our residents give us the green light for fire department funding and opportunities. All it takes is one individual to screw that up, and your fire department will be begging on their hands and knees for anything. Firefighters are supposed to be role models for the community. People trust us, and our job is to serve them.

One wrong move on social media can erode public trust completely. If the public feels like you've turned your back on them, they'll feel abandoned, and that's not bad only for you; it's damaging to your department and the entire fire service. We can't afford for the public to lose their faith in us.

The first person hearing about public backlash to your post won't be you; it'll be the chief. Depending on your department's policies, this could be the end of your dream job. Before posting, ask yourself if it's worth losing the career you've worked so hard for. Personally, I like to keep things low-key. I don't need the world knowing everything about me—because the less that people know, the less drama I have to deal with.

Keep your department information and title off your social media. If the "People you may know" identify you as "Firefighter/EMT So-and-So at [Insert] Department" in your biographical details, the senior members will notice and roll their eyes; they've already formed an opinion about you before you even know it. If you decide to switch departments one day, that impression may follow you. Sure, it's different once you're an officer or chief, but as a probie, everyone's watching your every move. The court of opinion can hold you back.

The Positives

Social media, if you're smart about how you use it, can still be one of the greatest tools to connect with like-minded individuals, share insights, and give back to the fire service. My accounts (@TheSaltyRookie on Instagram and TikTok) are taking off. Instagram is my space to show the realities, lessons, and mindset that come with this job. It's helped me build real, meaningful connections with fellow firefighters, including some of the biggest names in the fire service. On TikTok, I share my faith journey and focus on spreading positivity. No matter how big or small your platform is, you have the power to impact others.

Exposure is a double-edged sword, though. If you don't use it responsibly, it will come back to bite you. I've learned that what you post, how you represent yourself, and the message you put out to the world all matter. So if you're going to be present on social media, use it to lift others up, to represent your calling with integrity, and to stay true to who you are. In that way, you can make it work for you, instead of against you.

Think Twice

Any time a firefighter gets mentioned in the news, especially with something negative, it's not just your name on the line, it's your title. Get arrested for driving under the influence (DUI), and within 24 hours, everywhere you look—newspapers, online forums, you name it—will be "Firefighter So-and-So arrested for DUI yesterday."

When you took the oath, you didn't sign up for just the job; you swore to uphold a standard. You made promises to your brothers and sisters in this service. One mistake could follow you the rest of your career, or even be career ending if you're not careful.

So, before you decide to engage with someone online or presume that it's a good idea to get political and state your opinions, think again. I highly advise against doing so. From the moment you take the oath, you're no longer just you; for the rest of your career, you're Firefighter So-and-So. Every action reflects on your title, for good or ill. Act accordingly. Keeping your job and personal life separate will spare a lot of headaches down the road.

Facing My Reality

For many years, I carried the weight of a reputation I had created through my own poor choices on social media. Back in high school, I posted controversial political opinions and beliefs that, looking back, were harmful, offensive, and uncalled for. At the time, I didn't know better. I was heavily influenced by what I'd been taught at home, where political views were constantly pushed on me and held over my head. I hadn't developed the maturity to think for myself, so I mirrored people who should never have been my examples.

Those choices turned people against me. I faced relentless bullying, received death threats, and became the target of rumors that made me out to be worse than I was, even though I had set the stage. The truth is, I hurt people. I didn't think about how my words and actions affected others, and I regret that deeply. I let my ignorance, stubbornness, and immaturity speak louder than kindness, and it cost me.

From 2015 to 2018, I was the kind of person I wouldn't want to work alongside today. I carried an ignorant attitude that said, "I don't care. They can deal with it," and I see now how toxic that mindset was. I wish I could go back and undo the damage, but I can't. What I can do is take responsibility and show that I've learned. My mistakes humbled me. They forced me to grow, and they shaped me into the person I am today.

To my high school class, I am truly sorry for what I said, how I acted, and any hurt I caused. I deserved the fallout I faced. While an apology won't erase the past, I hope it shows that I've grown and changed. If there's one lesson I can pass on, it's this: Be careful what you put into the world. Think before you speak or post. Words have power, and they can either lift people up or tear them down. I learned that the hard way, and I hope others can learn from my mistakes without repeating them.

Key Takeaways

Social media is permanent. Once it's out there, you can't take it back. People will remember what you say and do online.

Don't assume that consequences won't happen to you. No one is immune. One bad post can destroy your reputation—and even your career.

Think before you post. Ask yourself, *Does this reflect who I truly want to be*? If not, don't post it.

Own your mistakes. Growth comes from taking responsibility, not making excuses.

Social media can be your greatest tool or your worst enemy. How you use it is entirely up to you.

Your words have power. They can either lift others up or tear them down. Choose wisely.

Don't let stubbornness cost you everything. Learn from others' experiences so you don't have to suffer the same consequences.

Silence is sometimes the best response. You don't have to weigh in on everything.

Final Thoughts

Learn from me: Don't assume, don't post carelessly, and don't think you're the exception. I never thought it would happen to me until it did. This wasn't my proudest chapter, and I'm not telling you this to preach; I'm telling you because I lived it. As hypocritical as it might sound, I'm advising you not to do what I once did. I had to learn the hard way, and that's what gives me something worth teaching. What separates me from the cowards is that I own my mistakes.

Social media can be a gray area, and I took it too far until it bit me. I won't make that mistake again, and I hope you won't either. Don't let social media be the reason your career ends before it even begins.

Reflection Prompts

- ✠ What is your department's social media policy, and what are the consequences for violating that policy?

- ✠ Have you been or witnessed someone being reprimanded for their social media posts? What was the outcome?

27

GEAR ESSENTIALS

February 10, 2025, 0400 hours

We're dispatched to a fire at a church, immediately after the sixth run of the night, and it is the worst call of my career. This shift has been a nightmare from hell, and I haven't once laid down in my nice warm bunk.

What is sleep? The night is loud, chaotic, and a never-ending battle. First on scene, type 3 construction, with no smoke or flame showing—I'll be investigating and assuming command until the ladder gets here. Pack on, mask connected, Halligan tool in hand, and a 360-degree size-up. Ladder 1 arrives and takes over incident command.

We enter the chapel, with nothing showing until we get into the auditorium. Thick smoke fills the balcony. "Stretch a hoseline, and tag the hydrant." I acknowledge the command and grab the most appropriate handline. But then I hear barked back at me, "Why'd you grab that line?" I swear my order was to grab a hoseline. I mumble to myself, "Well, this is the one I grabbed, so this is the one we're using." Just within earshot, I could hear him say, "Whatever."

Continuing on with my task and efficiently moving the 1 3/4-inch attack line, I set it, forget it, and go to the next assignment: tagging the hydrant to establish a water source. The process also requires grabbing the high-rise pack. Guess who's still fiddling with the water source? That's right: O'Donnell. Grabbing the hydrant bag and HAV, I struggle as the minutes progress, but I just can't get this water source established for the life of me.

My internal monologue is chiding me: *Come on, O'Donnell. Get it done, because if this church goes up in flames, you're in for a hell of a "Dear Chief" letter.* I continue to fiddle with this hydrant setup. Plan A is not working. What happens when plan A doesn't work? Go to plan B. And then plan C. When plan C doesn't work, you better think quickly!

The captain calls for a first-alarm assignment. What felt like hours was less than 10 minutes. By the time mutual aid arrives, I'm still struggling. It's 15°F out; everything is frozen, the ground is an ice rink, yet I'm drenched in sweat. I'm fighting this HAV setup like our lives depended on it. In fact, it did—because if we see flames and my crew are inside while I'm outside, someone's getting hurt or not coming home. In the heat of the moment, I'm anxiously playing orders back in my head.

I stroll to the fire department connection (FDC). The building connects to your water so that you don't have to haul hose up stairs. A standpipe feeds every floor. Unfortunately, I don't have the tool to get the covers off for it—and all I would have needed was a flat-head screwdriver. So I radio to the officer in charge, "I need assistance with the FDC. I'm still trying to establish a water source." The response comes back: "We're all tied up. You're gonna have to figure it out." I'm feeling overwhelmed at this point, and my backup plans have not been established. I didn't follow one of my biggest personal rules, to always have a plan. Shame on me! I should have radioed more proactively.

Suddenly, I see the first-arriving mutual aid crew pull up, and I ask for their help: "I need help with the FDC. I don't have the tool to get the covers off." While he journeyed over to take care of that issue, I went back to double-checking my HAV value setup. A minute later, the driver walks back over and lets me know, "The FDC is all set and connected." Relieved, I asked him to take a look at my setup. He studies it for a moment, then shrugs and says, "Yep, I have nothing." There's still no water established, much less the hose connection for the FDC.

I am panicking at this point, telling myself to start over. The time lost is unacceptable to me, because I have a job that I'm paid to do, that I train on regularly. Something's not working, and I need to think fast because I have the most important job on the fireground right now: getting water to the fire (even though there actually wasn't a fire in this case). Either I sit here and pout, without a solution, or I go back to basic skills and figure out a way to get the job done.

I try to disconnect all the pieces, but they're frozen. *Are you kidding me?* Of all things, my connections are frozen. I have no other tools that I can whip out to make this happen. *Think quick, O'Donnell.* I put up a last fight with this frozen solid connection, using all of my might—and hallelujah, I get that bugger to release!

With the FDC issue finally addressed, I attached the HAV value to the steamer of the hydrant. Why didn't I begin with that—after all, it's how I train? Instead, I set up some never-before-done hydrant tag—and I sure won't do that again.

Why are we trying new things on the fireground, O'Donnell? Couldn't tell you. I turn the hydrant back on, and at last we have water established. Finally, I let the officer in charge know that water has been established. I'm annoyed because I looked like a total newbie who has never seen a pump panel before. At this point, all I can do is laugh.

All crews are tied up. I'm standing there waiting, with *Jeopardy*'s 30-second song rifling through my head. While I had been assigned to the pump panel, the interior crews were looking for the source of the smoke, which wound up being an hour-long "Where's Waldo" search, eventually finding the culprit: a crusty heating, venting, and air-conditioning unit about to ignite. Tasks are assigned to all committed companies; next up: positive pressure ventilation.

Shortly after, the smoke lifts from the chapel, although the aroma is still heavily present. The shift officer begins relieving all mutual aid companies but one, while our two crews tidy up the scene. The chaos slowly gives way to calm, and I take the opportunity to pull one of my crew to the side: "Can you show me what I did wrong—because nothing went right, and it kept getting worse as the task went on?" "Sure thing," he replies.

I had succumbed to tunnel vision, the biggest mistake you can make on the fireground—which can kill you. I made a minor mistake, leaving behind a simple tool, that could have been easily resolved had I taken a step back to think about what I was doing. And if I had that one tool, I wouldn't have been in this predicament to begin with; I would have had at least one task completed. I had too much going on in my head after the worst call of my career, and my performance on this chapel alarm was affected as a result.

I won't do that again, and I learned from my mistake. That's what this job requires—always learning. Take it all in; take a step back and start over if you have to. If you aren't learning, then you aren't trying. You learn every day, no matter how much time you have on the job. The moment you hold the know-it-all attitude will be the day you slam headfirst into a wall. Your probie year won't be the only one where you make mistakes.

A mutual aid crew member walks up to me during the clean up: "What happened over there? Was the valve not working or something broken?" "Well, plan A didn't work, so I went to plan B. Plan B didn't work, so I went to plan C. All hell broke loose, and then I went to plan Z, which worked." As I explained my actions, I said—partly to myself and partly to the crew member—"Don't look new."

Create a set of minimal tools for your turnout gear, keep these on your person whenever riding on a truck. Eventually, with experience, you can add more tools if you can accept having heavier gear. There's an adage that when you say "I've never had to use this tool" (in however many years on the job) it ends up happening that you do—and when you least expect it. With my FDC issue, if I had the tool in the first place, then I wouldn't have had to ask for help. But since I held the mindset that "I won't need that tool," it happened to me in a situation that could have had a very poor outcome.

Simple tools can be essential for survival on a call. If you don't carry a minimal set at all times, you might end up in a life-changing situation, confessing, "I should have listened to my officer when he told me to get those tools."

Must-Haves and Why

I'm going to list the must-haves for your first year. Be prepared. It's always better to have something you don't need than to need something you don't have:

- *1-inch webbing.* There's a debate about whether you need 20 or 25 feet. Go with the bigger number and get 25 feet. It's better to have more than you need than not enough. You can always tie knots to shorten it if necessary.
- *Carabiners.* Think of carabiners like heavy-duty clips. I carry two at all times—one on my webbing and one on the waist harness of my turnout pants. Carabiners are crucial in firefighting, just like in rock climbing, for securing yourself or others in life-saving situations. My personal preference: I lock the one on my webbing but leave the one on my waist unlocked. Here's why:
    - For the webbing: If I have to rescue someone, I don't want the carabiner slipping and unlocking midsave.

   - For myself: If I need to bail out of a situation and visibility is low, I don't want to be fumbling with unlocking a carabiner. I know the one on my waist is ready to go, with no guessing involved.
   - A tip to store your webbing: Take the two free ends of your webbing and tie a water knot, with safety knots at each end. Attach a carabiner to the water knot. Roll the webbing, starting from the closed end—farthest from the water knot—until you reach the knot with the carabiner. Then, shove the rolled webbing into an EMS glove. This keeps the webbing secure, prevents it from unraveling, and protects it from abrasions. Plus, EMS gloves are easy to tear away if you need the webbing fast.
- *Wedges.* Carry at least two: one aluminum and one wooden. If you lose one, you've got a backup. They're great for more than just forcible entry: Use them as door chocks, or even to pry open a vehicle door when you need a *purchase* point (the spot where you start prying the door open). I stick mine in my helmet band for easy access.
- *EMS gloves.* Always have a couple of pairs on you. From medical calls to hydraulic line bursts, they protect your hands and gear. For example, hydraulic oil will eat away at your gloves, so you don't want that on your structure gloves. Tip: If a hydraulic line bursts, clean up with absorbent clay (Speedi Dri or

cat litter), and if the oil gets on your windshield, brake cleaner will get rid of it. (Brake cleaner also dries out poison ivy, but it hurts like hell. So I wouldn't recommend skin contact—even though I do it anyway.)

- *Multitool and pocketknife.* You never know when you'll need these. Get one with a window punch and seatbelt cutter to save space on your gear. Trust me, someone will ask to borrow these at some point.
- *3502 Key.* If your community has elevators, this key is a must for elevator rescues. Make sure your truck carries one or ask your officer to get one added.
- *Safety glasses (ZR87+).* Your eyesight should be a priority. Anytime you're dealing with bodily fluids or broken glass—for example, a car accident—wear them. It takes only one little piece of glass to ruin your vision for good.
- *Extrication/mechanic's gloves.* These are great for motor vehicle accidents or regular training when you want to save your turnout gloves from wear and tear.
- *Spare flashlight.* If your main light dies, you'll be glad you have a backup. Some light is better than no light in a dark building.
- *Ear protection.* Protect your hearing. Over time, constant exposure to loud noises will damage your hearing, so use ear protection when you can.

- *Screwdrivers, ¼-inch and 5⁄16-inch.* These are the two most common sizes for getting into rooftop units. Carry a 6-in-1 screwdriver with different-sized flat and Phillips heads to cover your bases.

Real-World Example: Using the 2:1 System in a Downed Firefighter Situation

Let's say you have a downed firefighter in a Mayday situation. Now, picture this: The downed firefighter, unconscious by a window, is the biggest person in your department. You couldn't lift them on your own—and even with help, it would be a monumental task. You need more than brute strength; you need the right tools and techniques. This is where the 2:1 setup comes in.

A 2:1 system is designed to help you capture and hold every bit of progress you make. Each time you move the downed firefighter forward, the system locks that gain in place, preventing them from sliding back. The beauty of the 2:1 system is that once it's set up, it reduces the effort needed to maintain your position, letting you focus on control and steady advancement rather than raw strength.

First, you need to check if a ground ladder has already been set up at the window. If not, radio for one. Once the ladder is in place, the next step is to get your rope and carabiners ready. Not everyone on scene will have the tools or

FIG. 27–1. Two views of a completed 2:1 setup: (left) progress capture—2:1 raise; (right) rescue load—firefighter down.

know-how to rig a 2:1 system, but you will because you're prepared.

Here's how to set it up:

1. *Form the figure eight on a bight.* Start by tying a figure eight on a bight at the end of your rope. This creates a secure loop, which will serve as the anchor point for your carabiner setup. Don't forget to add a safety knot to make sure everything is as secure as possible.

2. *Create the tensionless hitch.* Next, attach your rope to the ladder using a tensionless hitch. This knot is simple but effective—it relies on friction, rather than a traditional knot, to hold the rope securely in place. Wrap

the rope four or five times around a ladder rung at the highest point of the window, where the firefighter will be lowered from. The great thing about the tensionless hitch is that it won't slip under pressure, so it's perfect for situations where you need to rely on the strength of the rope itself. After wrapping the tensionless hitch, make sure that the loop with your figure eight knot is positioned just above it.

3. *Attach the carabiner.* With the figure eight knot and tensionless hitch in place, now attach your first carabiner to the loop formed by the figure eight knot. Then clip the second carabiner onto the first carabiner. At this point, you should have the tensionless hitch securing the rope to the ladder, and the carabiners are connected to the rope's anchor point.

4. *Create the 2:1 setup.* Take the running end of the rope (the part that's being fed from the rope bag) and form a bight (a small loop) at the end. Feed this bight through the first carabiner but not the second one. Once the bight is through, unclip the second carabiner from the first and clip it to the bight. Now, this carabiner and bight will be sent down to the firefighter below.

5. *Lower the rope.* Before you send the rope down, call out "Rope, rope, rope" to alert the ground crew. Drop the rope bag behind the ladder rung (never in front of it). Once it hits the ground, feed the remainder of the rope under the first rung and back to where the ladder

touches the ground. This will allow you to safely manage the rope as you begin the rescue.

6. *Capture your progress.* Here's the magic of the 2:1 system: It captures every bit of progress you make. Let's say you manage to move the downed firefighter an inch. The system holds that inch, preventing any backward movement, so you never lose ground. Every pull moves the firefighter safely toward the window for extraction. Once everything is in place, you and your team can use this system to slowly and safely lower the downed firefighter to the ground, controlling the descent rather than risking a catastrophic fall. Because you were prepared and had two carabiners on your person, you made this rescue possible when no one else could. You've now become the reason why that downed firefighter didn't burn alive.

Right Tool, Right Place

There are so many opinions and instructions as to where you should keep your tools. Put them wherever makes the most sense. For example, if it's a lifeline tool, have it so it's easily accessible when you have your air pack on. No way is the perfect way. The coolest thing about this job is that every firefighter has a different way of doing things, and there is no right or wrong way as long as it doesn't create a safety issue.

Make your tools essential for the most common calls you go on. If you're in a city with high-rise buildings, carry a bail-out kit. If you're in a district with a high volume of motor vehicle accidents, carry a window punch with a seat-belt cutter. If you're in a district surrounded by water, have a floating key chain so if you drop your keys they don't sink into the abyss. If you're in a community-oriented small town, have plastic fire helmets with you everywhere you go—kids love them! Be efficient and set yourself up for success.

Key Takeaways

Preparation matters. The tools you choose to carry can make or break your success on the fireground. Never assume you won't need something, because that's exactly when you will.

Tunnel vision is dangerous. Always take a step back when things go wrong. Reset, reassess, and go back to basics. The ability to adapt saves lives.

Train like you play. Fancy setups and untested ideas have no place in an emergency. Stick with what you know, and train until it's second nature.

Plan for failure. When plan A and plan B fall apart, you need to be ready with plan C—and be willing to pivot to plan Z if that's what it takes.

Ask for help but don't rely on it. Mutual aid is there for a reason, but your priority should always be to have the skills, tools, and mindset to handle the task yourself.

Carry the essentials. Webbing, carabiners, wedges, EMS gloves, multitools, screwdrivers, and eye protection should always be part of your gear. One simple tool could be the difference between success and disaster.

Final Thoughts

No matter how much time you have on the job, the fireground will always find a way to humble you. The key is how you respond when it does. Don't aim to look good; aim to be good. Carry what you need, train like your life depends on it (because it does), and never stop learning. The day you think you've got it all figured out is the day the job teaches you otherwise. Be ready, stay humble, and remember that it's better to have it and not need it than to need it and not have it.

Reflection Prompts

- ✠ What tools do you carry?

- ✠ What is the most frequent tool needed for calls in your jurisdiction?

28

THE GRIND

April 2, 2025

Once again, 8 months into paramedic school, I've beaten the odds. I've surpassed every expectation of failure and won out over every bet placed against me. I'm thriving.

When I look back at myself as that baby EMT who knew nothing, barely got by, and struggled to find her way, it almost feels unreal that I'm here. Now, standing in paramedic school, I've proven to myself that I belong here. There's still so much to learn, but I've found what works for me.

Today was the first day of clinicals, and I'm confident I'm going to be a fine paramedic one day. I may not have a perfect 4.0, but I've worked my way up to a 95% semester average. Trial and error got me this far, and I'm ready to grow into the woman I'm meant to be—not just a firefighter/EMT, but a firefighter/future-paramedic/author. This is only the beginning. Things are looking up from here, and I'm excited!

What's ahead is bigger than anything I've faced so far: mastering cardiology, conquering the rest of paramedic school, earning my license, and stepping fully into the role

I've worked so hard to earn. I know the road won't get easier, but I'm ready for it. Every challenge, every setback, every triumph has prepared me for what's next.

Maybe that's what makes standing here today so surreal—because getting to this point didn't happen overnight. It took grit, patience, and a refusal to quit. And that's exactly how I'm going to see this through.

June 16, 2025

Today was another day intubating patients in the operating room. We started cardiology a month ago, and it's one of the hardest challenges I've ever faced. My grades have taken a hit—not because I'm not trying, but because I'm human and I'm not perfect. What I've done so far is more impressive than I even gave myself credit for.

August 9, 2025

Today is my official last day of paramedic school. I DID IT!

It's been a hell of a ride—finding out so much about myself and realizing what I'm truly capable of. When the odds are stacked against you, when everyone says you can't, and you prove you can—that changes everything. I've gained confidence I never had before. I believe in myself now in a way I didn't know was possible.

Earlier diary entries mentioned my grades being good—but what I omitted was the fight behind them: keeping a high enough GPA to land on the Dean's List all four semesters of my program. That meant 40, 50, even 60 hours a week of studying, sketching anatomical diagrams, reading until my eyes blurred, hopping between coffee shops, classrooms, and kitchen tables for 6 or 7 hours straight—wringing out every drop of focus.

Because my brain works differently, I've had to work three times harder than "most" people. Notes on notes, highlights everywhere, nights spent screaming into pillows, crying on the floor, doubting myself to the core—you name it, I lived it.

When I almost quit in June, I thought, *If I give up now—after coming this far—I'd prove everyone right. I'd show I was a quitter who didn't keep her word.* But I fought. I fought the fatigue, the doubts, the impostor syndrome, so I could stand here today and tell you this:

Yes—you can do it. Any vision, any dream, anything you decide is worth it—all it takes is refusing to quit. As Marilyn vos Savant wisely said, "Being defeated is often a temporary condition. Giving up is what makes it permanent."

Failure isn't the end—it's often just the turning point. It means you're reaching, not running from what matters.

And here's the full-circle moment: Paramedic school didn't just train me—it launched me. That writing test I took to qualify for this school? It planted the seed. The long nights, the clinicals, the pages of notes, the resilience it took—all of it prepared me to be the author I'm becoming.

So as much as this was a journey of struggle, it also made me who I was meant to be. It threw me exactly where I needed to be—and pointed me toward what's next. That girl who never expected to amount to anything? She built herself from the ground up. Now, she stands here—a firefighter, an aspiring paramedic, and a writer—ready to tell her story. Maybe she'll inspire you to tell your story, too.

Do It Uncomfortable

There's a voice inside all of us that whispers every reason we can't, every reason we shouldn't even try. That voice is relentless, playing on your insecurities, digging at your scars, striking exactly where it hurts. It tells you that you're not enough, that you're bound to fail, so why bother? I know that voice too well. It haunted me for years, echoing in every corner of my life, tearing down anything I attempted to build. But that voice lies. The only thing standing between where you are and where you want to be is the courage to tell it to stop.

Looking back, I wonder how I ever got out of my own way. I was the queen of self-doubt, ruling through procrastination. High school was a blur—just getting by, content to stay in the background, finishing without fuss. Art was my one strength. I was good at it—really good. But I buried that part of me, convinced that making a future out of it was nothing more than a fantasy. Instead, I did things the uncomfortable way.

When I graduated, college felt like a trap. The thought of more classrooms and lectures made my skin crawl. I toyed with business and marketing, somehow getting accepted into a few schools despite my learning disabilities. But deep down, I knew that path wasn't for me. I didn't belong behind a desk. I craved the outdoors, where I could work with my hands and see the results of my effort.

Trial and Error

September 2017 through June 2018

During my senior year, I worked at a car dealership, moving up from the garage to customer service. I was even offered a sales role at 17 years old, but the idea of being stuck inside under fluorescent lights, pushing people to buy, felt like a prison. I wanted dirt under my nails and sun on my face. That's when I applied to a landscape company. Quickly, I was leading my own crew, tending more than two dozen properties a day. The pride of laying down perfect, clean stripes on a lawn; the hum of the mower; and the smell of fresh-cut grass? These were real, providing something I could cling to in a world that felt uncertain.

July 2018 through November 2019

Then came the Harley dealership. For a kid who grew up racing motorcycles, it felt like stepping into a dream. But that dream was short-lived: I was laid off before I even had the chance to find my footing.

So, I went back to landscaping, chasing that sense of purpose, until I applied to and got the job with the tree company the next year. I started with no clue, not even knowing how to hold a chain saw, let alone fell a tree without putting everyone in danger. But I threw myself into it. Fifteen months later, I was promoted to foreperson, leading my own crew. I earned my commercial driver's license by 20 years old—not without failing first, but failure only fueled me. I became hooked on the adrenaline of chasing goals and proving my doubts wrong.

May 2021

While still working for the tree company, I joined a volunteer fire department. The second I stepped into that firehouse, I knew this was it—the camaraderie, the adrenaline, the purpose. Nothing had ever hit me like that before. I wanted a full-time role with every fiber of my being. But here's the truth: As much as I wanted it, I didn't always show it.

Looking back, I can acknowledge where I fell short as a volunteer. I wasn't as motivated as I could have been. I let myself be held back by insecurity, self-doubt, and that familiar voice telling me I was not good enough.

Wanting isn't enough. At that stage, I hadn't learned how to match my wants with the work. I kept grinding at the tree company, trying to juggle both worlds. The fire in me hadn't fully caught yet. It took time; it took failure and doubt; it took realizing that if I truly wanted this career, I had to show up for it in ways I hadn't before. Every step from the tree crew to the firehouse prepared me for what came next. I learned to fight to earn my place—to have the courage to do it the uncomfortable way and to keep going when the voice in my head said I couldn't.

July 14, 2023

When a full-time role finally opened at the department I'd set my heart on, it was Joshua Sterls who gave me the heads-up. Sterls had been my mentor from the start and is one of the smartest firefighter/paramedics I know. He took me under his wing when I joined as a volunteer and helped me find my footing in this world. In 2022, he had landed the same full-time position I now dreamed of, and he showed me the way.

The job required a paramedic license, but I didn't even have my EMT yet. I was still an emergency first responder and had already failed the National Registry of Emergency Medical Technicians (NREMT) certification examination twice. Each failure fed the voice that said I wasn't good enough. I was ready to walk away. But Sterls wouldn't let that happen: "What's the worst that can happen? They say no? You'll never know if you don't try." That was all the encouragement I needed. Our friendship, built over years of working side-by-side, became a lifeline in that moment. Josh believed in me before I believed in myself. Without him, I wouldn't have applied, wouldn't have taken that chance, and wouldn't have written this book.

June 1, 2023

I applied even though I didn't meet the requirements, with no paramedic license. I knew I was up against seasoned candidates, with experience. Getting a spot in one of the most competitive departments around would be like winning the lottery. The waiting was brutal. Days blurred together—filled with anxious, sleepless nights and tangled in self-doubt wondering if I'd even get a call back.

July 18, 2023

I stared at the screen, heart pounding so hard I could feel it in my throat: I had received an email invitation to interview. How was this possible? I didn't have my EMT or my paramedic license, which were both requirements for the job. I hadn't touched EMT material in over a year. And now I had just 9 days to prepare for an interview that could change my entire life.

Panic hit at first, but something stronger rose up: a determination I'd never felt before. I registered for my third go at the NREMT examination (if I failed again, remedial training would be required). The only available date was 2 days before the interview. Those 9 days became a blur of relentless focus. I skipped meals, barely slept, and studied like my future depended on it—because it did.

July 27, 2023

The day of the examination came and went quickly. When the test shut off after 70 questions, my stomach dropped. I'd been here before, and it hadn't ended well. I walked out convinced that I had failed.

July 29, 2023, 0530 hours

I woke up to find an email: "Congratulations on earning your NREMT." Against all the odds and despite every whisper of doubt, I did it. Just in time.

Prepared to Win

August 8, 2025

I walked into that panel interview ready. Four captains sat across from me. When they asked about my EMT status, I told them the truth: I'd earned it 2 days earlier. I laid it all on the table, sharing my story of the failures and the fight to be there. They liked what they heard. I later found out that out of 30 candidates, I was the only one who had submitted a cover letter and resume. While the process is overall very tight-lipped, the feedback received was that I came prepared.

After the interview, I sent a thank-you email to the panel. That same evening, the chief called for a second interview. As excited as I was, I stayed focused. Nothing was guaranteed. There were still other candidates on the list. The second round went well, and soon came instructions for the third and final interview. Deep down, I felt that this was my position. But I kept it to myself, knowing the process wasn't over yet.

That moment changed everything. They didn't just hire me; they believed in me. I wasn't just another name on a list. I was the first woman they'd hired in almost 15 years. That carried weight, and I knew I had a lot to prove. But if they saw that I was capable, it meant that I was. For the first time, I felt truly seen. They recognized the potential I had struggled to see in myself. That belief became my fuel, the reason

Key Takeaways

Success doesn't happen alone. It's built on courage and persistence, with support from the people who believe in you when you struggle to believe in yourself.

Failure isn't the end. Facing your failure teaches you how to get back up stronger and more determined.

Preparation matters. Sometimes it's the small details, like submitting that cover letter or showing up ready, that set you apart.

The voice of doubt never fully goes away, but you can learn to rise above it.

The journey isn't about a single win. Earning the job wasn't the finish line; it was the start of becoming who you were meant to be.

I got my life together, and the reason I set out to build a career grounded in integrity, honor, and resilience.

This was my main character moment, and I was ready for whatever challenge came next. Now, deep into paramedic school, I carry that same determination that got me here in the first place. Every setback, every doubt, every victory led me to this point. That job offer wasn't the finish line. It was just the beginning of a bigger journey—one I'm still on today, fighting every day to become the firefighter, paramedic, and person I was meant to be.

Final Thoughts

The path to your goals will never be perfect—and it doesn't have to be. You're going to face setbacks, doubts, and moments when walking away feels easier than pushing forward. That's normal. What matters is what you do next.

Success isn't about never failing. It's about how you respond when things don't go as planned. Every obstacle is an opportunity to learn, adjust, and come back stronger. The process of chasing what you want will test your resilience, but that builds true strength.

Remind yourself that you don't have to do it all alone. Let people in. Seek out those who believe in you when you're struggling to believe in yourself. Leaning on others isn't weakness, it's how you grow.

What I've learned isn't just for me; it's fuel to help the next person chasing a dream they're scared to reach for. My hope is that my journey reminds others that no matter where you start, you have the power to rise—and eventually lift someone else along the way.

Finally, remember that earning a title or reaching a milestone isn't the end of the road. It's the beginning. The real work starts after the door opens, when you decide how you'll show up, who you'll become, and how you'll honor the opportunity you fought for. Wherever you are on your journey, know this: The odds may not always be in your favor, but with hard work, humility, and heart, you can defy them.

Reflection Prompts

- ✠ What inspired you to join the fire service?

- ✠ Would you do anything different in the steps you took to get here? Why or why not?

- ✠ Who have been some of your biggest supporters? How did they help you?

29

DON'T RUSH TO LEAD

I feel strongly about this topic, and the reason why boils down to one word: responsibility. This job isn't just a job. It's life-and-death. It's showing up on the worst day of someone's life and making sure everyone—civilians and your own crew—goes home in the end. That's the weight that comes with the fire service. If you're going to carry that weight, you had better be ready for it.

June 2025

In 4 years on the job, I've observed a lot. I've experienced firsthand what happens when people rush through the ranks, chasing titles before they've put in the work to truly understand the job. I've learned the qualities that make a strong leader—not on paper, but in person.

In my opinion, the Massachusetts Firefighting Academy has it right. They require at least 3 years as a full-time firefighter before you can even think about applying to be support staff. However, we allow people to take officer and instructor courses after just a few months on the job. A standard should be set for minimum amount of time in the field before being eligible to lead.

I've also seen what it looks like when leadership is earned—built on experience, integrity, and the confidence that comes only from doing and failing, then learning and doing it better the next time. Too many people confuse passing tests with being prepared. But books and certificates don't fight fire; firefighters do. Good leaders are forged on the fireground, not in a classroom.

Leadership isn't something that magically clicks into place the day you get promoted. It starts the moment you walk through the firehouse doors—from the jump seat, from mastering the basics, from helping the next guy up. If you're not ready to lead from the jump seat, you sure aren't ready to lead from the front.

The Reality

A common concern I've heard from several friends in other departments is the rush to promote. In some places, members can move up in just 3 years if they meet certain requirements on paper. But passing a written test doesn't always mean someone is fully prepared for the job. I've met firefighters who were promoted quickly but still struggled with basic operational skills—because real competence comes from experience, repetition, and mastering the basics over time, not just checking boxes.

The truth is that rushing to promote in this job gets people hurt or killed. Anyone can pass a test, but can you handle the heat when it really matters? Leadership isn't something you can just flip on when you suddenly find yourself with a shiny new rank. It starts the moment you raise your hand and take the oath, even before you walk through the firehouse doors.

Leadership Begins from the Jump Seat

You don't have to be an officer to lead. Leadership begins when you're still figuring the job out. Even 6 months in, you can lead. You show leadership when you help the new guy who's even greener than you. You lead when you're the one who steps up, gets people off their rear ends, and gets them out there to train or try something new. It's not easy, but learning how to lead from the jump seat prepares you for the long haul, regardless whether your promotion is 10 years away or just around the corner. Don't wait for a promotion to start leading—and if you do, don't expect to gain respect from anyone.

The biggest lesson I've learned is to ask for constructive criticism and learn from your mistakes. If you don't know the job inside and out, you're not ready to lead. It takes only one person to shift the culture in a department, for better or worse. There are leaders who inspire and leaders who do the bare minimum. Don't fall into the trap of thinking, *I've been promoted, so everyone has to respect me.*

Sure, anyone above you in rank deserves respect, but there's a huge difference between respect for your title and respect for your work ethic. Be serious about the job, not about chasing rank. Be the one who sets the standard. When you get comfortable in the firehouse, you start getting complacent. Search online on "firefighter complacency" and read the horror stories. If those stories don't put anxiety in the pit of your stomach, maybe you're in the wrong profession. Leadership isn't about the title. It's about the actions you take and the example you set every single day.

Own the Moment

Never forget that as a firefighter, you're in the best seat in the truck. You're the one on the nozzle, getting hands-on experience, fighting fire head-on, and doing what most people only dream of.

When you move up in rank, though, everything changes. You're not the one charging into the flames anymore, because your role is to make sure those in the seat you used to occupy are doing exactly what needs to be done in the right way. It's your responsibility to guide your crew, keep them focused, and above all, keep them safe. You'll be the one ensuring the nozzle is advancing, the line is laid correctly and the crew is following orders to the letter, with no shortcuts. They need to know that you're there to hold them accountable, not to hold their hands. The stakes are too high for mistakes when lives are on the line. There's no room for sloppiness or hesitation.

The outcomes on this job are irreversible. Sometimes that means accepting that you've killed one of your guys as a consequence of not doing your job well enough. That's on you. Your job is to lead your crew by example, then watch their backs, not your own.

Don't let the burden of leadership discourage you. Plenty of seasoned firefighters still say this is the best job in the world. What keeps us from being miserable as leaders is knowing when to have fun versus when to call the shots on the fireground. Leadership means responsibility; it doesn't mean becoming a broken record. Personally, I know some officers who still don't feel ready and live with many regrets.

Key Takeaways

Leadership isn't about rank. It's about responsibility, accountability, and setting the standard through your actions, not your title.

Experience matters. No test or certificate can replace time on the floor, mastering the basics, and learning through doing.

Lead from where you are. You don't have to wait for a promotion to start leading. Leadership begins with how you show up every day.

Complacency kills. The moment you get comfortable is the moment you put yourself and your crew at risk. Stay sharp and stay humble.

Fast-tracking leads to failure. Rushing into leadership roles without enough experience doesn't hurt you alone—it puts everyone around you in danger.

Final Thoughts

Leadership is earned, not handed out. There are no shortcuts, easy paths, or substitutes for doing the work. Be the firefighter who others can count on, not just because of what's on your collar, but because of how you carry yourself, how you train, and how you treat your crew. The best leaders I've known never stopped learning and mastering the basics, and never forgot the responsibility that comes with the job. Don't rush to lead; instead, rush to be better today than you were yesterday. The rest will fall into place when the time is right.

Reflection Prompts

- ✠ Have you seen anyone promoted before they were ready? What impact did that have on the team?

- ✠ Think of a leader you admire. What specific thing do they do (not just say) that makes you respect them?

30

SURVIVING TO THRIVING

Reflecting on my first year as a firefighter, I'd call it survival of the fittest. Not because of the fires or the dangers I faced on the job, but because of everything else that comes with being the new kid on the block. The long hours, the relentless pressure to perform, and the unspoken expectations to prove yourself are enough to burn anyone out. The challenge is not just getting through the shifts, but learning to thrive in the chaos, to laugh at your mistakes, and to keep grinding when you're not sure if you belong.

This is my testimony of survival, overcoming adversity, finding the resilience to push in the darkest times, and coming out stronger on the other side even when the odds were against me. Many individuals judge me when I walk into a room because of how I present myself—strong, driven, and passionate—when they should take a minute to learn about me before forming an impression. A dismissive mindset reflects how judgmental humans can be. However, sentiments like "She thinks she's better than everyone," "She's a people pleaser," "She's way too confident," and "She's cocky" are not "me" issues; they reflect the character of those passing judgment.

Now let's pull back the curtain to show you the human being behind the gear. This isn't a story just about being a probie firefighter, but about finding myself while navigating one of the most physically demanding and rewarding careers. Firefighting can consume your identity, but our personal journeys and triumphs define us, too.

Welcome to my raw, real, vulnerable, unfiltered, and unapologetic testimony. In this chapter, you'll be taken on an interesting ride through the darkest and most incredible events of my life. These trials, tribulations, and gritty moments shaped me into the woman I am today. This is the first time I've shared in depth about the truths behind the mask in my life; even though some people already know part of my story, they don't know who I really am or why I work as hard as I do.

My story will change others' perceptions—positive or negative—of who I am, but I'm completely OK with that. The opinions that others form won't ever change how I think about myself because I know where I came from and the work it took to get here. This is another reason why I never presume myself to be above anyone; I always treat others the way I want to be treated because I never thought I'd be where I am today. I've been given a once in a lifetime opportunity, and I'm so grateful for it—it still doesn't feel real to me.

Welcome to my story of coming out stronger on the other side. You are going to know everything about me by the end. If I can change one person's life, that's all that matters to me—not money, fame, or recognition, but the impact of these words and emotions.

Blood Doesn't Mean Family

September 2003

I was molested by an older male relative at 3 years old. The case never stuck. For years, I was told, "You're crazy, you could never remember anything that young."

I remember clear as day the touching when no one was around. In the early hours of the night, he walked up to my crib and began touching and tickling my genitals, playing with them. It was dark, cold, and I hadn't a clue what was going on, other than that his was a familiar face.

When I got a little older, I mentioned to my mother how he "was playing with me and touched me down here." I understood little then about how wrong it was, but it turned into a big case, involving child protective services and the district attorney. My mother spoke on my behalf, and next thing I knew, I found myself at 5 years old in the district attorney's office being asked to demonstrate where I was assaulted. Having to share in this way while extremely vulnerable—essentially forced to point at my genitals to a random room of strangers—was the most uncomfortable thing I've ever experienced. I didn't understand what was going on at the time, but years later, when I realized what the big deal was, it disgusted me. That messed me up for a very long time—it's not easy to recover from, that's for sure.

2004

In the chaos of the house I grew up in, the sound of my parents' drunken screaming was as normal as the ticking of a clock. Alcoholism was in my family on both sides. It echoed through every corner of our home until my parents divorced. They each fought their own battles, and I'm not here to unpack their past history. What followed, though, shaped the next chapter of my life in ways I never once imagined.

After the divorce, my mother took custody of me. She decided that the best way to deal with me was to keep me away from my father until I was 12 years old. I was too young to understand her perspective and reasoning, right or wrong, but I did know that I was being kept from my father. And I loved my dad. Having a broken family so young was not ideal, and I wouldn't wish it on any one.

People love to throw around phrases like "Family is everything," but they rarely talk about when family is the source of your deepest wounds. When it's not love that binds you, but control, manipulation, and silence. My father's side of the family made that reality impossible to ignore. My dad spent his entire life being treated as an afterthought. While his sisters—Martha and Ginny—were handed college tuition, financial support, and every safety net, he was given nothing: no help, no investment, just expectations without resources. He had to grind through life while they walked across bridges that were built for them. And when he started a family of his own, the same lack of respect bled into the way they treated us, his daughters.

We were never good enough—never included in the same way. We were the "other side" of the family, the ones who didn't measure up. Because we weren't made of money, that made us "lesser" in the eyes of my dad's family, and they made sure we knew it. When I was a kid, visiting my grandparents' house felt more like a test than a reunion. Aunt Martha would stab my elbows with a fork at the dinner table if they so much as grazed the edge. If I said "Oh my God," she'd threaten to slap me for using the Lord's name in vain. These rebukes always occurred when my dad wasn't around. It wasn't discipline; it was a power trip. And no one ever stood up for me.

Years later, after working 13 grueling days in a row —17-hour shifts in the freezing cold, cleaning up after a brutal storm—I showed up for Christmas at my grandparents' house. I was exhausted, but I wanted to be there while my grandparents were still alive. I sat down on the couch for a moment to rest my eyes, and Martha shook me awake and pinched me aggressively, snapping that I was "rude" for trying to sleep. She said I needed to "stay awake and be present," as if dragging myself there in that condition wasn't enough. She worked in a temperature-controlled office, while I had been outside, in snow and ice, breaking my back just to show up—and it still wasn't enough. That moment wasn't unique. It was just another reminder of how they always saw us as less.

My sister, Tiffany, who lived with and cared for our grandparents while managing a full-time job, was treated no better. When our grandmother passed away, our aunts swept in and moved our grandfather into a nursing home hours away and closer to them—more convenient for their schedule, yet nearly impossible for us to reach with our jobs and responsibilities. They sold the house Tiffany was living in and gave her minimal notice to pack up and find new housing. "It's not our problem," they told her. Then they changed his will, silently but strategically. Everything of value—financial and sentimental—went to them, while we were left with crumbs. The car, the house, the stocks, the keepsakes were gone, stripped away like we never mattered. It's been over 5 years since my sisters and I have heard a word from either of our aunts. And I don't care, because they showed us exactly who they were. But the damage they caused wasn't just about inheritance or distance. It started long before any of that.

After my parents divorced in 2004, my mom had full custody of me. This meant that my sisters, Tiffany and Danielle, lived separate lives, walking different paths, because we all share the same dad but they have a different mother. Even though I'm only their half-sister, that doesn't change how I feel about either of them—I love them both. I saw my dad on weekends; still, time was limited, and the dynamic was never balanced. He tried, but he was always fighting uphill. My mom controlled the schedule, the structure, the narrative, and he just had to take what he could get. It wasn't fair to him—and it wasn't fair to any of us. We all had a rough upbringing, with distinct but still very traumatic experiences being our shared history. While their stories aren't mine to tell, I carry deep empathy for what they endured, regardless of how much blood we share or how long it took to build a relationship. We were just kids trying to survive a world that didn't make room for us to be seen, heard, or understood.

Therapy, or a Test?

Under my mother's care, I was dragged to countless therapy sessions, each more frustrating and disheartening than the last. I was told I needed medications because I'd been labeled (unprofessionally, since it's not a diagnosis per se) a "problem child." They forced my mother to dose me with their prescribed medications. A diagnosis of attention deficit hyperactivity disorder followed, and I was prescribed Adderall and a couple other medications—because prescribing medications to a child who doesn't like their therapist was the answer for everything. That label of "problem child" stuck for nearly a decade, as I was forced to sit in those suffocating rooms with an adult who acted like they knew me but never once tried to understand. I was told that I wasn't smart enough, that I was weak minded, and that the only way to fix me was through prescriptions and therapy I didn't want.

All of this led to a surplus of issues in school. The moment I started the medications, several problems began: twitching, saying weird things, losing the ability to focus, and swearing (and I had no clue I was doing it). Teachers would pull me to the side of the room to ask politely what was wrong, while students would be more blunt: "Why are you doing that? You're weird." It got to the point that when I was about

11 years old, I remember passing out and waking up on the kitchen floor with my head split open, shortly after taking this medication. My father walked in and said, "We're done with these medications they put you on. No more, kiddo."

I'll never forget the therapist who made my childhood a living hell. She wasn't there to help me, having zero warmth in her eyes and no genuine care for my well-being. Just there to collect a paycheck, she went through the motions without concern for the damage she was doing to me. Every session felt like a punishment. Constant threats to my freedom, the one thing that mattered most to me, hung over my head like a dark cloud. For instance, I would lose my bicycle and the ability to play outside with my friends if I didn't talk during one of those miserable forced sessions.

I was just a kid who needed someone to believe in me, to see beyond the label of problem child. Instead, I was met with indifference, judgment, and manipulation. I'd never treat anyone, let alone a child, the way my therapist treated me.

Dear Mom

Early 2004

At 4 years old, I remember getting out of bed at 1 a.m. to use the bathroom. I nearly tripped over my mom, who was unconscious and overdosed on God knows what. This wasn't her first suicide attempt, and many more would follow. The image of her on the passed out on the floor is permanently etched into my mind and has haunted me since.

January 2013

The last time she tried to take her life, I was 12 years old. I remember that day vividly, because I found her note tucked away for my father to find. She blamed me. Being told that I was the reason my own mother didn't want to live shattered something in me. No child should ever have to carry that level of guilt, that intense of pain. Please, never put that kind of weight on a child's shoulders.

For the first month after the event, school was different for me. The staff had been informed of the circumstances and behaved honorably and with kindness. Still, I had unanswered questions: Was my mom alive? What was going on? And why was it my fault?

I was sick to my stomach, and I stopped eating. Being as young as I was, they didn't exactly tell me what was going on, but I could tell that things weren't looking promising for my mom. At one point, they told me that the probability of her dying was likely. For the next year, my life was in turmoil—moving to new schools and starting a new life with dad—and I had no thoughts about how my life would unfold into adulthood. Eventually, my mother made it out of the hospital alive after that last suicide attempt, and as difficult as it was, I made amends with her.

Facing the Darkness

January 24, 2013

The day I found mom overdosed, I didn't scream or panic. Instead, I quietly slipped the leftover opiates at her bedside into the pocket of my hoodie as I gathered my possessions to move into my father's house. I had already made up my mind and had my plan ready to go; 12 years old, I was already tired and numb. My mom had tried to kill herself and failed so many times. I was convinced I would follow through where she hadn't succeeded, because it was all I knew. I clung to those pills as my secret escape route.

Not a single soul knew I carried this heavy darkness. It was well hidden—behind forced smiles and silence. At the time, I thought it was normal to feel this way.

June 2015

My friend, Alana, found the opiates in my room and, being curious, asked what they were. I couldn't pretend anymore. I broke down, and without a word, she got up and flushed them, every last one. That moment shattered my last attempt at pretending I was fine. My secret was out, the weight of years spent hiding from the pain I never wanted to face.

I wouldn't be where I am today without Alana, who saved me by flushing those pills down the drain. We lost touch over a decade ago, but one day, I got a tap on my shoulder. It was Alana. She was at the hospital while I was on a transport. I found out she works there as a medical assistant. We reunited after 10 years—what a godsend.

For many years, I was left to reconcile with the way my mother had twisted everything, turning herself into the victim. Alone, I grappled with the darkness my mom had planted inside of me. Chills ran down my spine from the damage she caused, but ultimately, I chose to be better. I now understand that she was battling her own demons—ones I won't ever fully comprehend.

Mental illness takes many forms. I wasn't at fault for the choices my mother made. I was simply the scapegoat her mind created to make herself feel better. If I were to place blame on her, then multiple fingers would point back at me. In the end, it's not necessary to place blame on anyone.

Some people shouldn't be given the gift of becoming a parent. Unfortunately, children are an easy target. Nonetheless, it ignited the fire of my success story. I wouldn't have the passion or drive if it wasn't for my mom's mistakes. While I will never forget the wounds she caused, I have chosen to forgive her. She didn't choose to have a mental illness.

Being an advocate for mental health is deeply important to me. If I pitied my mom's hardships but then preached about forgiveness, that would make me a hypocrite. As my friends at Back of the Bay always say, "Do better." It's simple, but it's a reminder I carry with me every day—to choose growth over grudges and to keep moving forward.

Junk-Food Diet

Damage experienced firsthand as a kid can't be undone. My mother fed me a steady diet of junk food, and we frequented McDonald's. My childhood was fueled by Coca-Cola, chips, and unhealthy snacks. I was overweight for years and never looked like the other kids, which made me an easy target for teasing. They called me lazy, fat, and constantly reminded me that I didn't belong. The bullying I faced was relentless, especially for what I wore. I wore nonfeminine hand-me-down clothes, so a lot of people told me I looked like a man.

Children aren't born judgmental. They act out based on what they're taught from an early age. Biases are learned.

Please don't influence a child negatively. Instead, teach them the Golden Rule: "Treat others the way you want to be treated." In particular, I'll never forget one nasty girl who sneered at me in the bathroom and said, "Ew, why are you looking at yourself in the mirror? You're nothing to look at." Her words still echo my mind.

My classmates were unrelentingly brutal to me, leading me to act out in high school. Later on my ultimatum was "If you feel good, you look good," and tired of being fat, I developed a fitness regimen. This experience fed into my work ethic of always wanting to improve. I got tired of being an easy target, so I used the bully's teasing to fuel the fire of my motivation.

Toxic Relationships

The relationships that followed seemed like an extension of the pain I had already endured. I was abused by people who claimed to love me, beaten, cheated on, and manipulated. I lost another piece of myself, encountering narcissists who chipped away at whatever worth I had left, making me question if I'd ever be good enough for anyone. For 8 years, I drowned those thoughts in alcohol. Drinking my teens away, I tried to numb the pain, partying with people I thought were my friends and hoping that the drunken buzz would silence the echoes of my past: neglect from my mom, bullying, and the cruel words that defined my childhood.

August 17, 2018,

Just before my 18th birthday, I was drugged. When I woke up, I found that out a "friend," who I had trusted for more than 5 years, had raped me. I was too intoxicated to understand what was happening or to stop it. I had bruises and welts all over my body, in unthinkable areas. I kept telling myself, "There's no way this really happened." It wasn't until I had a urinary tract infection a couple days later that I acknowledged the real reason for the immense pain in my genitals.

September 18, 2018

A month later, I received a gut-wrenching message from the perpetrator:

> *I know you probably don't care about me or anything I'm about to say but I am sorry for that night. I wish that didn't happen, maybe we'd still be talking right now. I do regret it and I don't. It was awesome and it was great but I do regret doing it to you at the same time because I really do miss you Nichole. I am very sorry. I hope you do great in Florida and have a great life bye.*

It was true that he'd raped me, and he knew it. He carefully worded the message, though, so that there was no blame to place or evidence. I felt caught in a downward spiral. I had no idea who I was anymore, and the only way I knew to behave was to mirror the people around me—which meant numbing the pain with alcohol to finding some sense of belonging with other drinkers. None of those people should've been anyone I looked to for guidance. I was lost, mimicking the worst parts of others, and it almost broke me.

Dear Dad

September 2022

It all began with a phone call from my father, a conversation that would alter the course of my relationship with him for years and influence my career choice. He told me his girlfriend was on her way to the police station to file slander charges against me. For years, ever since they got together, there had been nothing but tension between us. I fought for my bond with my dad, hoping that he would choose our relationship over the conflict she created. Painfully, though, I was losing my dad to another woman after all that I had been through.

She turned our relationship into a competition for his attention, but in the end, he chose her. I wanted to believe that he wasn't really abandoning me, that he was just afraid of being alone. But when he called and said that she was going to ruin my life, everything inside me shifted. He told me I could kiss my firefighting career goodbye: "I hope you have a good lawyer because I'm not helping. You did this to yourself. I didn't raise you like this." Something fundamental had broken between us that could never be repaired, and a deep sense of betrayal resulted.

My heart was shattering, and I was grasping at anything to make it stop. I was sobbing, shouting, trying to make him understand how deeply hurt I was. But what came out was rage and desperation. I told him to go fuck himself, and he didn't even flinch.

He hit me where it hurt the most, wielding his words like knives: "You have a mental illness, just like your mother. You belong in the psych ward, just like her." He didn't care how much it hurt or how fragile I was at that moment.

I replied, "Then I'm going to kill myself. You'll have to live with that. You'll have to live with knowing you did this to me." My voice was shaking, but in that moment, I meant every word. It wasn't just anger, but pain so deep that I didn't know how else to react.

In response to that—no concern or panic in his voice—he called me stupid, like I was nothing. He didn't care that I was on the edge or hurting. Then, he hung up the phone.

I waited for him to call back, prove me wrong, to show me that deep down he gave a damn about me. But the silence was deafening: nothing, not even a call to the police for a well-being check.

The next day, every last piece of my life was stuffed into boxes, and I walked out the doors of the place I had called home. He didn't call or text. He didn't check to see if I was alive. Later on, he claimed he didn't remember saying any of it and blamed the alcohol he was drinking. But I remember every word of that phone call and the way it felt when my own blood father shrugged off my cry for help.

That was when I truly knew I was on my own for the rest of my life. The man who was supposed to protect me had turned his back. His girlfriend had won, leaving me standing in the wreckage. No one was coming to save me this time, so I had to save myself. On my own from that moment forward, this became the reality: At the end of the day, the only one who cares about you is yourself.

Our Relationship Through the Years

From the day I was born until I turned 16 years old, my dad was my rock, the one person who I could always count on. He supported me in everything, especially during my racing days, and he always had my back. We were close in a way that only people who really know me could understand. After my mom did what she did, my dad stepped in and did his best to fill the void. It wasn't perfect, but he did the absolute best he could, and he was good to me. For most of my life, my dad was my best friend—even when he'd had minimal visitation rights during my mother's initial custody. We had a bond that I thought was unbreakable.

Overnight, he became just another person who walked out of my life, leaving me to deal with the pain on my own. But I knew that I couldn't let this define me.

My dad's girlfriend took away the last person who mattered to me. Her weapon wasn't drugs, as it had been for my mom, but manipulation tactics. It hurt deeply, but the truth of this loss was undeniable.

Since then, I've been gradually rebuilding a steady relationship with my dad—even though it won't be the same. In the end, he's still my dad, and I love him regardless. Things weren't ideal for me growing up, but he tried—and I'll always give him the credit for that. I'm proud of him for stepping in when he didn't have to. Despite the challenges I faced, he gave me the best life he could with the hands we were dealt.

Pain into Purpose

I spent the next year rebuilding myself, one step at a time. I became sober and pushed myself to grow physically, mentally, and emotionally. I became the woman who would have protected the girl I had been. As the pain smoldered, I had to find forgiveness, for myself and the ones who had hurt me the most. I couldn't change what had happened, but I could choose to let go of the anger and forgive. Accepting what had happened—not with resentment but with understanding—I forgave both my father and his girlfriend, not just for them, but for myself. I stopped blaming myself for the hurt and began to appreciate what the pain that I endured had taught me.

This chapter isn't about retribution or bitterness; it's about finding forgiveness within. Everyone makes mistakes, and even though the choices others made hurt me, I've found peace and I forgive them. The scars are still there, but they remind me of how far I've come. I'm thankful for the entire journey, because they led me to discover forgiveness within myself.

The dark days are behind me now, and they won't return. Every hardship became fuel for my ambition, which pushed me to become the woman I'm proud to be. I am thankful for what happened because it shaped me into who I am today. If things hadn't unfolded the way they did, I wouldn't be where I am now. That pain pushed me harder than anything else, driving me to prove my strength to myself. Ultimately, the pain forced me to grow.

February 2025

I made amends recently with my dad and his girlfriend, and casting aside resentment is one of the best decisions I've ever made. Now that we are civil, she and I sat down for coffee, which was a humbling and peaceful encounter. I enjoyed my time and truly found forgiveness within. You can't change what happened, but you can make the best of it. My dad's girlfriend may not be blood, but she is still family and deserves to be treated with respect as an equal.

Everyone on this earth makes mistakes. If I couldn't forgive my father or his girlfriend, then it would mean I couldn't forgive myself. By forgiving, I found happiness with those around me. I look forward to more coffee mornings and dinners with them!

A Calling

May 2021

This wasn't just a career choice for me; it was a calling. To step into someone's life on their worst day and to be the one offering compassion, comfort, and closure when it's needed most demands empathy, strength, and the ability to truly listen. This role calls for greatness, and every day I push myself to meet that standard. No matter how much I give, though, a part of me wonders if it's enough. In grade school, I was labeled lazy, so here I'm going to shatter that label. Every hardship has shaped my resilience, so now I can guide others through hardships.

My true beginning in the fire service was a quiet start in 2021, when I took a step through the doors of a small-town call department, asking for a chance to be a part of something greater than myself. I didn't know exactly what I was searching for; I just knew I needed direction, purpose, and a reason to believe that I was capable of more than the voice inside my head allowed me to believe. Walking in with baggage, I was loud, nervous, guarded, and trying to prove I belonged while battling my demons internally. Something about that firehouse felt real. It welcomed me like a place I'd been searching for my whole life but didn't know existed. For the first time in years, I felt safe—like I could drop my guard just a little. There was a mix of acceptance and challenge, like they were saying, "You're one of us now, but you'd better keep up."

It wasn't just being part of a crew—it was feeling seen. Every call, every training drill, every laugh in the kitchen chipped away at the voice in my head telling me I didn't belong. Slowly, that voice got quieter. In its place came something new—pride, purpose, and the unshakable sense that I was exactly where I was meant to be.

June 2022

I graduated from the call/volunteer Massachusetts Firefighting Academy. Getting my certificate lit a fire I didn't know I had. It wasn't just a piece of paper—everything shifted. It told me I belonged. That I could do hard things. That I had earned my place. And it reminded me that I was only just getting started.

That same drive carried me into paramedic school where the challenges were bigger, the hours longer, and the stakes higher. The road wasn't smooth. There were nights I wanted to quit, days when the weight of the workload felt impossible, and moments when I doubted I had what it took. But the lessons I learned in that firehouse and at the academy never left me: Show up, keep going, and earn it every single day.

Now, as I stand here at the end of my paramedic program, I can see how it all connects. That first step through the firehouse doors set the tone. The academy built the foundation. Paramedic school tested it. And every step—every win, every setback—was preparing me for this moment.

This isn't just the end of one chapter. It's the continuation of a story that started the day I found where I belonged—and it's proof that I'm exactly where I'm meant to be.

Entering the Career Side

October 23, 2023

I transitioned to the career side of the fire service. The learning curve steepened, the expectations rose, and the weight of the job was like nothing I had experienced before. I graduated from the full-time Massachusetts Firefighting Academy in January 2024, and 12 days later, I was operating my first structure fire. No easing into it, just boots on the ground, mask on, and go. I had to rely on my training, my instincts, and the people beside me.

Every shift after that kept challenging me, stretching me, forcing me to become more, not just as a firefighter but also as a person. It was through that intensity that I realized this job was never about being perfect; it's about being present and rising to the occasion when others are falling apart. It's about trusting your crew and yourself and doing the right thing when no one's watching.

One of the most defining features of this chapter in my life was the presence of a strong leader: Captain Tony believed in me when I couldn't yet believe in myself. He has been a leader who didn't let me coast and didn't let me fall alone. He raised the bar and showed me I could meet it. His leadership shaped me, teaching me to lead with accountability and heart and reminding me that great leaders don't just demand respect but earn it through consistency, honesty, and presence. Because of his influence, I carry myself differently now, knowing the kind of firefighter, teammate, and future leader I want to be. This is something that can't be taught in a classroom.

Degrees Don't Define Your Capabilities

With my history of academic struggles, I turned my disabilities into my drive to thrive—not just as a firefighter, but in life. Not being a great student in school doesn't define you, I promise.

I stopped letting other people's opinions decide my future. I wasn't living for my parents' expectations or anyone else's approval. I did what I wanted to do. Don't ever let someone hold a career over your head—do what you want to do. It's OK to try something, finish school, and then realize it's not for you. It happens—trust me. Sometimes you have to try something to see if it works, and if it doesn't, you start over. It's a process, but it's worth the experience.

I remember rushing through my senior year of high school, so excited to graduate. But once you're out those doors, reality hits—you have bills to pay and responsibilities waiting for you. I wasn't book smart, so I took the hands-on route. For me, trade schools were the way to go; I didn't want years of debt.

Try everything. You don't have to have life figured out by 20, 30, or even 40. Every step, every attempt, every restart—it all adds up to who you become and the skills you carry forward. Good things come to those who work for them, not to those who wait for them.

School isn't for everyone. We all have our niches and things we just can't see ourselves doing—and that's OK. A degree doesn't define your capabilities in life. It's OK to not be a doctor, even if your family expects it of you. Do what you want to do. You never know until you try. If you have an interest in a topic, explore it. If you end up not liking it, then at least you tried it, and you'll never have those lingering "What would have happened if I tried the thing?" questions nagging you.

Full Circle

I think back to that loud, nervous probie who walked into the firehouse in 2021. She was unsure of herself and had no idea where it would all lead. She didn't know the nights she'd cry, the fires she'd fight, or the strength she'd discover. She didn't know how many times she'd be tested, or how many times she'd get back up. She didn't know she'd become Private O'Donnell. She didn't know she'd finally find her voice, her peace, and her purpose.

If I could meet that version of myself at her call/volunteer graduation (fig. 30–1), I'd sit beside her, take her hand, and say, "If only you knew where this would all lead. Because it led here to this moment, this version of me."

FIG. 30–1. Massachusetts Firefighting Academy call/volunteer graduation, June 24, 2022 (Photograph courtesy of Kaitlynn Chaivralotti)

Key Takeaways

One of life's biggest lessons is learning to forgive, because if we don't, you get nowhere. Being an angry person isn't fun. You'll feel much better when you just accept what happened, learn from it, and move on—no matter how much it hurts. No amount of frustration can change what has already happened and the damage that has been done. You can't change the outcome, but you can handle how you react to it.

As my captain Anthony Arventos taught me, "Don't stress yourself out over something you have no control over. You'll drive yourself into the ground doing that." Developing that mindset doesn't happen overnight. This journey has taught me more than I can put into words, but I'd still like to leave you with a few truths:

- You don't need to have it all figured out to take the first step.
- Your past does not define your potential.
- Mistakes are part of the process. Own them.
- Ask questions. Stay humble. Keep growing.
- Take care of your mental health; it matters more than you know.
- Check in on your people, especially the ones who smile the most.
- Great leaders change lives. Follow them, learn from them, and become one.
- Forgiveness is freedom.
- You are not broken. You are becoming.
- If you keep showing up, even when it's hard, you will rise—and you will be rewarded.

Final Thoughts

If you've made it to this page, you have walked through a piece of my life: the fire, the failures, the growth, the grit. Thank you for staying with me until the end. That means more than I can say. I'm proud of you for making it this far, both in this book and in your own story. This was my way of making a difference in the world, and letting you know that you're not alone, not now or ever. If even one sentence in here made you feel seen, stronger, or a little less alone, then it was worth every page.

People close to me would tell you that I don't shut up. Truthfully, though, I've lived a life that demanded a voice. For years, I brushed off the pain, laughed off the fear, and pushed forward as if nothing bothered me. Writing this book cracked that wide open. It forced me to feel everything I spent so long trying to numb. I'm not perfect. I've made a thousand mistakes, yet this chapter of my life is closing. Not with a period but with a comma, because another book is coming.

Thank you for giving this salty probie from Massachusetts a chance and reading my story.

FIG. 30–2. Massachusetts Firefighting Academy Career Recruit Graduation, January 27, 2024 (Springfield, MA) (Photograph courtesy of Jared Berg)

Reflection Prompts

- What hardships have you had to overcome in your life, and how did they mold you into the person you are today?

- What were your best and worst experiences as a probie?

SIGNATURES AND NOTES

INDEX